The COMPLEAT FARMER

A Compendium of Do-It-Yourself, Tried and True Practices for the Farm, Garden & Household

Compiled by The Main Street Press

Introduction by Charles van Ravenswaay

Main Street/Universe Books

New York

Copyright © 1975 by The Main Street Press

First Edition

Second Printing, 1976

Library of Congress Catalog Card Number 75-10934

SBN 0-87663-928-7

Published by Universe Books,
381 Park Avenue South, New York
10016. Produced by The Main
Street Press, 42 Main Street, Clinton,
New Jersey 08809.

Cover design by Helen Iranyi

Photographic reproduction by Fran Levinson

Printed in the United States of America

Contents

Foreword

Did the tomato you bought at the supermarket this morning taste like sawdust this evening? Did the "fresh" cucumber (the one *without* the yellow blotches) cost twice the price of a packet of cucumber seeds? And was it so thickly wax-covered that it fairly slipped off the check-out counter? Did the plastic broom you bought two weeks ago fall apart the second time you used it? Did the stitching on your store-bought shirt or jacket or blouse disintegrate after its first wash? Did last month's electric bill and telephone bill and tax bill and grocery bill make you wish that you could catch the first train leaving the twentieth century?

If you've had a hard time finding that train — or *any* train for that matter — then this book is for you.

The Compleat Farmer is an indispensable guide to good country living. It is, as its subtitle suggests, "a compendium of do-it-yourself, tried-and true practices for the farm, garden, and household." But it is more than that. It is the sage, sound, salient advice of the nineteenth-century American farmer and his wife, selected, edited, and arranged for its practical use *today*.

Culled from more than fifty years of the pages of *The American Agriculturist*, the leading farm journal of the previous century, *The Compleat Farmer* is a repository of the useful, the practical, and the inventive. It speaks eloquently of an era in which the majority of Americans lived and worked on small homesteads and of the pioneering days before farming became big business. Whether it be step-by-step directions for building a stone fence, an explanation of the uses to which a windmill can be put, or the recipe for an inexpensive, but nourishing vegetable soup, these serviceable, interesting ideas from America's past speak directly to America's present. All are immediately applicable to a society bogged down with energy and cost problems, wanting to cut back, wanting to live a simpler and less costly life, but not knowing how or where to begin.

Let *The Compleat Farmer* show you how.

The instructions in this book will enable you to move from field and woods to vegetable garden and flower bed, from orchard and barn to root cellar and springhouse, from kitchen range and home canner to quilting frame and hand-made furniture. With this book in hand you can clear your land and fence it properly, raise ducks and sheep, keep honeybees, or process maple sugar. Or you can grow brussels sprouts and cauliflower, celery and leeks, horseradish and kohlrabi, okra and lima beans, melons and quince, rhubarb and sweet potatoes, shallots and tomatoes, blueberries and strawberries, peaches and cranberries. Or you can enjoy the old-fashioned delights of hollyhock and lavender, sweet pea and snapdragon. You can even learn the secret of making your own essence of violets or explore the intricacies of the Victorian floral vocabulary in which each flower represented a particular sentiment or human emotion.

With *The Compleat Farmer* as your guide, you can master the best way to store apples, potatoes, and cabbages, or learn to prune vineyard and rose garden, or prepare green tomato pickle in your own homemade cider vinegar. You can bake all manner of breads — from corn to ginger to potato — and prepare a mock-pumpkin pie made from turnips. And you can enjoy a lesson in middle-American table manners: "Do not touch your hair while at table, nor pick your teeth — *and above all do not suck them.*"

Complete directions, illustrated with charming nineteenth-century woodcuts and engravings, enable you to make your own hammock and summer awnings and to build fences from cedar brush, buckthorn, holly, osage orange, wooden poles, board, wire, or stone. And on cold, winter evenings, you can turn your thoughts to making homemade soap and candles, baskets and brooms, rag carpets and straw mats, and even an easy chair from an old barrel.

The Compleat Farmer is not a catalogue. It does not merely direct you to sources of information. It *contains* the information. It offers the step-by-step farm, garden, and household tips and hints so valuable and necessary if you are to reap and enjoy the fruits of the earth. In its practical, straight-from-the-shoulder manner, it can assist you in your return to the soil — and in your flight from the world of plastic hair rollers and transistor radios. Its methods are all organic, all natural, all conceived and practiced in a day when America was still green.

Even if you own no land, or if you are only thinking of buying land, *The Compleat Farmer* is a wonderful book to daydream by. It offers a fascinating view of nineteenth-century America — before the age of electricity and before mass production and merchandising virtually eliminated home industry and the native-American craft tradition. And its language reflects the very spirit of nineteenth-century rural America — hard-nosed, but oddly lyrical; practical, but imaginatively inventive.

The Compleat Farmer, borrowing its title from Henry Peacham's *The Compleat Gentleman* and from Izaak Walton's *The Compleat Angler*, can complement your efforts to become an accomplished, or "compleat," agriculturist. It is great fun to read — and to use.

Compiling *The Compleat Farmer* has been a complex but rewarding experience. From thousands of articles, enough to assemble a dozen volumes, those particularly contemporary and useful were selected. Each was edited and reset so that the text would be eminently readable and attractive to the eye. Illustrations both instructive and amusing were chosen to complement rather than to decorate the text.

Charles van Ravenswaay, director of the Henry Francis du Pont Winterthur Museum, first suggested such a volume. An authority in the decorative arts, he has, nonetheless, lost none of his enthusiasm for the very practical, thoughtful, and inventive genius of America's nineteenth-century farming community. His Introduction to *The Compleat Farmer* follows.

AMERICAN AGRICULTURIST.

Designed to improve all Classes interested in Soil Culture

AGRICULTURE IS THE MOST HEALTHFUL, THE MOST USEFUL, AND THE MOST NOBLE EMPLOYMENT OF MAN —WASHINGTON

ORANGE JUDD, A. M., } **ESTABLISHED IN 1842.** { **$1.00 PER ANNUM, IN ADVANCE.**
EDITOR AND PROPRIETOR. } { **SINGLE NUMBERS 10 CENTS.**

VOL. XVIII.—No, 2.] NEW-YORK, FEBRUARY, 1859. [NEW SERIES—No. 145.

Introduction

October.

During the nineteenth century an agricultural revolution transformed farm practices and farm life throughout America. Improved grains, fruit, and livestock were introduced; scientific knowledge of farm methods found increasing acceptance; and laborsaving farm equipment came into general use. As the nation's population grew westward, better roads, along with steamboats and railroads, opened new markets to farming communities, encouraging greater productivity.

Many of those who witnessed this transformation in rural life had lived as children on small, self-sufficient farms which since colonial times had produced little more than small crops, scruffy livestock, and much hard work. Many had been a part of the westward migration from rocky hillside farms in New England or southern plantations grown sterile by tobacco culture. Change and promise marked the century, and those who had been a part of it recalled in their old age its wonder and excitement.

During this agricultural revolution, farm journals informed, interpreted, and assessed each facet of these innovations, encouraging change by visions of a better future. Hundreds of journals came into existence during the century, and in time a majority of the nation's farmers were their subscribers. Many were poorly edited and soon disappeared. Those managed by men familiar with the new agriculture, and who understood the needs and aspirations of their readers, often had long and useful lives.

In April, 1842, when more than thirty farm journals were in circulation, Anthony B. and Richard L. Allen launched *The American Agriculturist* in New York City, hoping through this means to help improve farming, particularly by improving livestock. These brothers were natives of Massachusetts, well educated, and apparently with comfortable means. As farmers in western New York they had become interested in stock breeding and their experiences and studies had made them familiar with farm needs generally. Anthony was active in importing purebred livestock from England. Richard was more of a student and literary man. A third brother, Lewis F. Allen, became a frequent contributor to the new magazine with articles on rural architecture and other subjects. Although the journal was of considerable quality, it failed to attract a strong following, perhaps because of its conventional format and lack of breadth and vitality. Readers then, as now, wanted facts but they also wanted something more.

Discouraged by the indifferent public reception and the mounting losses, and undecided about how they might salvage their effort, the brothers turned to retailing and manufacturing agricultural implements as a more effective, and more profitable, way to improve agricultural practices. Then in one of those accidental meetings that seem preordained, A. B. Allen met a remarkable man with the improbable name of Orange Judd. He had just finished graduate studies in agricultural science at Yale and was in search of a career. Judd was hired as associate editor and set about with energy and imagination to "make a paper that people would want, and . . . to let people know of it." His masthead announced that the periodical was "designed to improve all classes interested in soil culture," and quoted Washington's statement that "agriculture is the most healthful, the most useful, and the most noble employment of man." In 1856, when he purchased A. B. Allen's remaining interest in the journal, the subscription list was 812. He eventually increased it to 160,000.

Orange Judd's life was the Horatio Alger story of a poor farm boy overcoming seemingly impossible obstacles. He was one of eleven children in the family of a farmer near Niagara Falls, New York. Undaunted by his limited opportunities, he left home and supported himself while attending Wesleyan University at Middletown, Connecticut, and later, Yale. His family background gave him a knowledge of the plight of the small landowner and a compassion for their problems. Valuable, too, was his understanding of their conservatism and unwillingness to accept new ideas and book learning. His courses at Yale had introduced him to the new agricultural science and brought with it a vision of how it could benefit all American farmers. That vision he pursued throughout his life with unfailing optimism and considerable courage.

As an editor, Judd believed that even the most scientific information should be presented in terms any reader could understand. For someone less skilled the results would have been tedious or condescending, but Judd made agricultural writing an art. This took much time and thought, he admitted, but it was essential for the "easy comprehension of *the people*." And it was the people he sought to reach. He considered his subscribers his friends and their interests his own.

Under Judd's management *The American Agriculturist* grew from twenty-four to thirty-six pages, printed in clear type on folio sheets of good paper with many excellent engravings. Of interest to the farmer were articles and brief notes on every conceivable aspect of farm operations. Correspondents reported on the state of agriculture in various parts of the nation. His readers contributed comments and information. One section was devoted to the interests of the farm wife. Judd encouraged the beautification of farm lawns and the introduction of new plant material. Often he wrote nostalgically about the beauties of old-fashioned flowers and gardens and was among the first to urge the cultivation of native plants. Farm children also had their own section with short stories, riddles, and games. Not least among the interesting features were the advertisements which Judd carefully screened and edited. Through this effort he became aware of the swindlers preying upon the farm population and so incensed did he become by his discoveries that in 1868 he began a regular column, "Sundry Humbugs," to expose them. His vehemence, his vivid descriptions of crooked schemes, and his hints of indecent books and pamphlets, must have fascinated and titillated his readers.

Judd energetically sought new subscriptions by offering premiums. This became so popular that supplements were issued whose pictures and descriptions rivaled those of early mail-order catalogues. In 1858 a German-language edition of the journal was started to serve the growing immigrant population.

EVERY READER OF THIS JOURNAL IS INVITED TO EXAMINE THE NOTES GIVEN BELOW.

I.

Is there any doubt that it will be **useful** for every family, and every person, everywhere, to get the information presented in the pages of this Journal during a year—in type and engravings?

Intelligent, experienced, practical men, who know, and *know well*, what they talk and write about, are constantly engaged in investigating, collecting, sifting and condensing the *best* useful hints and suggestions to fill the columns.

Besides the Editors, many prominent men, worthy Instructors in Leading Institutions, and others foremost in pushing forward improvements in the industries of our country, send their best thoughts to these pages. Many thousands of dollars are expended on engravings which speak more plainly than words can do.

The reading matter and engravings in this Journal during a single year, would fill half a dozen Books costing an average of $1.50 each.

Is it possible for any one, whatever his occupation, to read a yearly volume without getting *many* thoughts, each one of which will in the end be very valuable, directly or indirectly?

American Agriculturist in German.

The AMERICAN AGRICULTURIST is published in both the English and German Languages. Both Editions are of Uniform size, and contain as nearly as possible the same Articles and Illustrations. The German Edition is furnished at the same rates as the English.

American Agriculturist.

(ISSUED IN BOTH ENGLISH AND GERMAN.)

A THOROUGH GOING, RELIABLE, and PRACTICAL Journal, devoted to the different departments of SOIL CULTURE—such as growing FIELD CROPS; ORCHARD and GARDEN FRUITS; GARDEN VEGETABLES and FLOWERS; TREES, PLANTS, and FLOWERS for the LAWN or YARD; IN-DOOR and OUT DOOR work around the DWELLING; care of DOMESTIC ANIMALS &c. &c.

☞ The matter of each number will be prepared mainly with reference to the month of issue and the paper will be promptly and regularly mailed at least one day before the beginning of the month.

A full CALENDAR OF OPERATIONS for the season is given every month.

FOUR to FIVE hundred or more, Illustrative ENGRAVINGS will appear in each volume.

Over SIX HUNDRED PLAIN, PRACTICAL, instructive articles will be given every year.

The Editors and Contributors are all PRACTICAL, WORKING MEN.

The teachings of the AGRICULTURIST are confined to no State or Territory, but are adapted to the wants of all sections of the country—it is, as its name indicates, truly AMERICAN IN ITS CHARACTER.

The German edition is of the same size and price the English, and contains all of its reading matter, and its numerous illustrative engravings.

TERMS—INVARIABLY IN ADVANCE.

One copy one year.................$1 00
Six copies one year..................5 00
Ten or more copies one year....80 cents each.

An extra copy to the person sending 15 or more names, at 80 cents each.

☞ In addition to the above rates: Postage to Canada 6 cents, to England and France 24 cents, to Germany 24 cents, and to Russia 72 cents *per annum*.

Delivery in New-York city and Brooklyn, 12 cents a year.

Postage anywhere in the United States and Territories must be paid by the subscriber, and is *only six cents a year*, if paid in advance at the office where received.

Subscriptions can begin Jan. 1st., July 1st., or at any other date if specially desired.

The paper is considered *paid* for whenever it is sent, and will be promptly discontinued when the time for which it is ordered expires.

All business and other communications should be addressed to the Editor and Proprietor,

ORANGE JUDD,
No. 189 Water st , New-York.

751 BROADWAY, NEW YORK.

THE GREAT RURAL PUBLISHING
HOUSE OF THE WORLD.

To encourage farmers to buy and study practical books he first advertised and reviewed new works as they appeared, then formed his own publishing company in 1865 which soon became the leader in its field. Its titles included a great variety of rural subjects, among them books on the design and construction of farmhouses, barns, and other buildings complete with plans, elevations, and cost estimates. These, and articles regularly carried in the journal, greatly influenced rural architecture throughout the nation.

For readers today the numerous "how to do it" articles carried regularly in the journal are endlessly fascinating. They were on practical subjects, with each step in the process clearly described. Often they were accompanied by illustrations. Many of these articles were supplied by Judd and his staff, but he also encouraged his readers to contribute information drawn from their own experience. They include accounts on how to build houses of hewn logs, sod, adobe, cement and gravel, and balloon-frame construction. Readers were told how to split rails and shingles; make bricks, baskets, brooms, and candles; thatch roofs; and build fences of stone, wood, sod, and wire. Directions were given farm wives on how to set a table (and on table manners), plan a convenient kitchen, clean house, and make decorative accessories. Making rugs from rags was explained, along with arranging bouquets and Christmas decorations. Even such mundane subjects as how to sweep carpets, mix whitewash and whiten walls, and churn, were seriously discussed.

Unusual in its time was the noncontroversial nature of the journal and its concern with farm interests in all parts of the expanding nation. Relgion, politics, and sectionalism were taboo. Traveling correspondents and staff artists reported on farm life throughout the country including the territories and new states of the Far West. Settlers frequently wrote vivid accounts of their experiences. Through these and other ways Judd sought to broaden the horizons of American farmers, and to create among his readers a community of interest.

Containing a great variety of Items, including many good Hints and Suggestions which we give in small type and condensed form for want of space elsewhere.

The Homestead Law.—The following letter from Speaker Grow, interpreting this law for a Kansas subscriber, will explain itself:

SPEAKER'S ROOM, June 20, '62.

Dear Sir.—No one person can take, under the Homestead Act, over a quarter section, and that he can take without any reference to how much land he owns, if he occupies and cultivates as required by law. But if he lives on less than a quarter section, he can take adjoining, provided that (which) he takes with what he has shall not exceed a quarter section. Your owning land in Kansas does not prevent your *settling on and cultivating* either an adjoining quarter section, or any other elsewhere.

Yours truly, G. A. GROW.

Settlements under the Homestead Act.—"J. C.," "H. W.," Melrose, will see by referring to our last issue that we advise the formation of companies, say fifty families or more, in making new settlements. This plants all the institutions of civilization in the wilderness at once. Get up such a company or join one already formed, if you move at all.

To Prevent Wall Paper from Fading.—A Connecticut subscriber recommends a lump of alum as big as hen's egg dissolved in 6 quarts of paste.

A Short Method with Night Caps.—"H. H. B.," Wis. Aunt Cinda's method is to tie a knot in each of the opposite corners of a handkerchief, about two inches from the end. Put it on the head one knot before, and the other behind, and tie the other two ends under the chin. Look in the glass and keep from laughing if you can. A shorter method still is to put a *not* before the whole kerchief instead of *knot* in the two ends. We eschew night caps and dogs. A clean head makes a clean pillow case. Soap and water at the roots of the hair are better than cotton at the other end. Keep the head cool.

Why Grease Bread Pans?—"Eunice," of Columbia Co., Wis., does not see the utility of the almost universal practice of greasing bread pans before baking. She does not do it. In the multitude of councillors there is wisdom. Who speaks? Many people bake the bread on the oven-bottom, sprinkled with ashes, and some to the ashes, or instead of them, add flour and carraway seeds. Such use no grease.

Rancid Butter is said to be cured by mixing soda with the cake or cookery in which the butter is mixed. We doubt the cure and still more the wholesomeness of the compound. Fresh butter, if you please.

Medicine for Summer Complaint. J. L. Holmes, Bristol Co., Mass., says, steep rhubarb root in water, add molasses, and boil to a syrup. If the patient is weak, add a teaspoonful of brandy to a teaspoonful of syrup, and give a teaspoonful 2 or 3 times a day.

Drying Sweet Corn for Succotash.—"J. F. H.," Lancaster, Pa. We know of no good oven or furnace expressly for this purpose, and one is hardly needed for preparing family supplies. Where a business is made of drying apples and other fruits they have an apparatus expressly for the purpose. For the family a bushel or two may be dried with the common stove without any difficulty. The corn is boiled slightly in its milky state, scraped or cut from the cob, dried a day or two upon sheets in a bright sun, carrying them in at night, and then finished off in a brick or common stove oven. No family should be without a full supply of this article. Succotash once a week is wholesome, economical, and "not bad to take."

Rain in England.—From tables kept in London, England, we find the amount of rain which fell in 1861 was 20.7 inches; the average fall there is 25.4 inches.

Judd's contributions to American agriculture extended far beyond the pages of his journal. He created the sorghum industry by distributing large quantities of free seed in 1856. Four years later he provided funds to set up the first agricultural experiment station in the United States at his alma mater, Wesleyan University. In another year he gave 3,000,000 packages of flower seeds to his readers. These and other innovative and useful acts helped make Orange Judd a household name throughout much of rural America.

During the first thirty years of its history, *The American Agriculturist* absorbed twenty-six other periodicals to become one of the strongest voices of the American farmer. By 1873 Orange Judd had created a remarkable publishing empire and in the process had grown wealthy, investing heavily in a railroad and real estate. The financial panic of that year ruined him. In 1879 he lost his remaining interest in the journal and in 1883 withdrew from the company. Although his editorial policies were continued by succeeding owners, the loss of his vital presence became quickly apparent.

Undismayed by these reverses, Judd, at sixty-two, moved to Chicago to manage *The Prairie Farmer.* In 1888 he and his sons bought the St. Paul *Farmer,* moved it to Chicago and renamed it the *Orange Judd Farmer,* which he published with increasing success until his death in 1892. *The American Agriculturist*, under successive owners and changes in direction, is still published in Ithaca, New York, a tribute to the enduring vision of the remarkable Mr. Judd.

Charles van Ravenswaay
Winterthur, Delaware

I Vegetable Gardening

"A DREAM THAT WAS NOT ALL A DREAM."

The Kitchen Garden

A good garden for raising vegetables and small fruits, is one of the most important appendages to a house. Indeed, a house in the country is not a home without it. It greatly promotes the comfort and health of one's family. Well, says quaint Dr. Deane, "I consider the kitchen garden of very considerable importance, as pot-herbs, salads, and roots of various kinds are useful in housekeeping. Having plenty of them at hand, a family will not be so likely to run into the error which is too common in this country—of eating flesh in too great a proportion for heatlh. Farmers as well as others, should have kitchen gardens; and they need not grudge the labor of tending them, which may be done at odd intervals of time, that would otherwise chance to be consumed in needless loitering."

The position of such a garden is a matter of considerable consequence. Probably the best aspect is a southern inclination; next to this a southwestern, or southeastern; and poorest of all a northern. It should, of course, be nigh the rear of the house, as to be easy of access from the kitchen; and as our old author writes, "not far away, lest being too much out of sight, it should be out of mind, and the necessary culture of it too much neglected."

A garden should be well fenced. For protection against thieves, nothing is better than a good thorn hedge, the thorn lucust being the most formidable. But for shelter from cold northwest winds, a high board fence or wall is better. This protection is very important where one wishes to raise early vegetables and tender fruits. It breaks off severe blasts, and gives a warm and summery air to the garden quite early in the spring. It may cause them to reflect the sunlight more powerfully, but it also makes them part with their heat faster.

The selection of a suitable soil is also a matter of great importance. By all means avoid a low, wet piece of ground; for though the brightest sun may shine upon it, and though you may heap the richest manures upon its surface, it will yet be unsuitable for a garden. Draining may help it, but can scarcely make it as warm and generous as one that is naturally dry. A light, mellow, turfy loam, neither very sandy, nor yet of a stiff clayey texture, is the quality most to be desired. It should be said, however, that skillful cultivation can modify an unfavorable soil much, making it lighter or heavier as it may need.

To prepare the ground for planting, it should be sub-soiled or trenched. If trenching, let it not be done in the thoughtless way sometimes practiced of throwing up the bottom earth, and burying the rich surface soil beneath it. In old and over-cropped gardens, it may answer to bring up a little of the lower soil annually, but as a general rule, it should be left at the bottom. The right way of trenching is this: begin on one side of your patch by digging a trench two feet wide and one spade deep, carrying off the dirt to the further side of the patch. Then go through this trench again, simply breaking up the subsoil to the depth of the spade, but not throwing it out. Now, fill up this trench with the good surface soil taken from the next por-

tion to be trenched, then dig the bottom of the second trench and go on as before. When the whole plot is trenched over in this way, the soil taken off from the first trench will fill up the last, and the work is then done. If the soil is poor, a coat of manure should be worked in as the trenching proceeds. If it is still and cold, give it a dressing of old leaves, sand, lime, or peat, or anything calculated to render it porous and friable. If too light, add a coating of ashes, clay, etc., to be thoroughly mixed with it.

The shape of a garden is not a matter of great importance, though the nearer it approaches to a square or parallelogram, the better. The internal arrangement is of more consequence. The quite common plan in most good gardens is substantially this: lay off a border from four to six feet wide, all around the outer side of the plot. Devote this primarily to vines and low shrubs. On the north side plant grapes, that they may have the full benefit of the sun. On the west set raspberries and blackberries. On the east put quinces and a few dwarf pears. On the south set currants and gooseberries of the various kinds which being of low growth will not materially shade the garden. A walk in front of this whole border may be from two to five feet wide, according to the size of the garden. So too, if space permit, a walk may be laid off through the middle of the garden, with a border on each side for dwarf pears or other small fruits. Grapes trained to poles may occupy a part of this border, if they are so managed as not to shade the other portions of the garden. If, for example, this middle walk runs east and west, the south border might contain grapes, because they would cast their shadow only across the walk; but the north border should not.

The remainder of the space may be laid off in squares for melons, squashes, cucumbers, cabbages, peas, and the like; or into beds for beets, onions, and other vegetables. (1850)

Muck—What is Muck?

In England, muck means manure. In *Pilgrim's Progress*, "the man with the muck-rake" was searching for good in the gutter's filth; but not finding that for which he searched. "Muck is money," is an English farmer's proverb, the meaning of which is clear enough. When agricultural writers in America talk about muck, they mean swamp muck, and by this, a substance of a peaty character, rich in humus, of a dark brown or nearly black color, consisting of the remains of plants which have undergone partial decomposition under the constant influence of water. This has no constant composition or appearance other than indicated.

In peat-beds, the true peat is often several feet deep, and there may be a good deal of similar material which is crumbly, more or less mixed with earth or sand, and unfit for fuel. Other deposits abound in which there is no peat fit to use as fuel, but with an abundance of other material useful to farmers, and properly enough called muck. This is black soil, at least half of which will burn away when dry. It often dries hard, like clay or bricks. It crumbles under the influence of frost and air, and often simply by drying. This substance, including all varieties of peat, is or may be made useful in every upland soil, indeed in every soil not of a peaty nature to begin with. It is often rich in nitrogen, the most costly ingredient of fertilizers, often contains phosphate of lime and other valuable ash ingredients. By its decom-

position in the soil, its absorbent action, its promotion of other decompositions and changes in the soil, its presence is always beneficial. Under some peculiar circumstances, these effects are hindered, probably by the presence of organic acids in the peat. To such peats and mucks the term "sour" is fitly applied by farmers. They may, however, be neutralized, or, so to speak,

sweetened by lime. Fresh-burnt lime rapidly absorbs water, and falls into a fine dry powder. The muck is spread in layers, a few inches in thickness, and lime in this form is spread thinly over it. It is not necessary to be accurate in regard to proportions, but best to be uniform. If the muck layer be about 4 or 5 inches thick, half a bushel of lime will be sufficient for a space of 10 feet by 10, or 100 square feet, and even be used for double that area. The muck being piled up in layers, each receiving its quota of lime, becomes changed—more easily pulverized and disintegrated, equally useful as an absorbent, and a superior ingredient of composts.

The muck or peat of some localities may be applied directly to the land, either fresh dug, if dry enough to haul, in which condition it is best to apply it in the autumn, so that it may become ameliorated by the frosts and thawings of winter, or after such weathering. Other kinds of muck cannot be used advantageously without composting with lime or manure, or with ashes, or some other active substance, while that of some localities, applied raw, is positively deleterious to the crops of the first year.

As a general rule, muck may be made most useful in ordinary farm operations by mixing it with manure from the stable, in the cow-yard, the pig-pens, or the sheep-yards, and it is safe to say that the addition of muck of good quality in this way may easily double or triple the value of the manure made upon the farm. That is, a yard capable of furnishing, under ordinary circumstances, 100 loads of manure, may be made to furnish twice or even three times as much, both in quantity and value. (1882)

Sweet Herbs

There are many things that are not food in themselves that help give a pleasing variety to foods; these are generally classed under the term of condiments. These may be mere flavorings, or may be added to help the digestion. With us the use of condiments is not a necessity; they are employed to give a pleasant flavor to articles of food. The plants called Sweet Herbs, comprise Sage, Thyme, Summer Savory, and Sweet Marjorams—sometimes Sweet Basil is added to the list, and Parsley, used mostly in the green state, may be classed here.

A bed or corner of the garden may be used to raise the young plants, which can be set out after early vegetables are off. They are best treated as annuals, though Sage and Thyme will live longer than one year. They grow most luxuriantly in late summer. The plants should be cut when in bloom, tied in small bunches, and hung up to dry; they may be preserved in this state, or when quite dry to be rubbed with the hands to a coarse powder, to be preserved in bottles in a convenient form. (1882)

The Sweet Basil

As a general thing, our gardens do not show a great variety of what are called "sweet herbs," or those aromatic plants used for flavoring in cooking. Sage, thyme, summer savory, and sweet marjoram, are all that are usually grown. The French housewife could not well do without tarragon and basil, the last being a favorite in Europe.

The common basil is *Ocimum Basilicum*, a much-branched, annual herb of the mint family, with minute white flowers in whorls. The name *Ocimum* is from the Greek, and refers to the odor of the plants, and *Basilicum* is from the Greek word for royal. In India the basil is regarded as one of the most sacred of herbs, which has the power to protect those who cultivate it from all evil, etc. We have seen it cultivated in old flower gardens for the sake of its perfume, under the name of "Bergamotte."

Its proper place, however, is in the kitchen garden, with thyme and similar herbs. The flavor of basil is quite unlike that of any other of the herbs used for seasoning, reminding one of cloves. It is esteemed by those who are acquainted with it for flavoring soups and salads. There is a restaurant in New York City celebrated for the excellence of the meat salads served at its lunch counter. This excellence is solely due to the use of basil in the dressing.

There are several varieties offered by the seedsmen, but there is no essential difference between them. Being tropical plants, the seeds should not be sown before the soil is warm. If wanted early, the seeds may be sown in a hot-bed, and when the weather is settled, the plants may be set out a foot apart. When in bloom, cut the stems near the ground, and hang the plants in an airy room to dry. When completely dry, rub between the hands, reject the large stems, and keep the coarsely powdered leaves in a tin box or glass jar, to be used as required. (1886)

About Capers

Years ago, when people made fewer conundrums than they do now, it used to be asked "when is a cook like a dancing master," the answer was, "when he cuts capers." It is probable that many of our readers have no idea at all of what a caper is, and would fail to see the point of the quibble. There are many trivial luxuries that are mainly confined to the large cities and the more wealthy, and without which farmer-folks can manage to live very comfortably.

These include many articles used in cooking that are not food, but only serve as

seasoning; for these in the aggregate, large sums are annually paid, and capers are among them. Capers come to us in odd looking, long and narrow wide-mouthed bottles, and look at a little distance like pickled peas; upon examination they will be found to be not perfectly round, but somewhat larger at one end than the other, and to have a short stem at the larger end. Ridges are seen upon the surface, and if one of these capers be carefully picked open it will be seen to be, what it really is, the bud of a flower.

The plant which produces capers is *Capparis spinosa*, a low straggling shrub which grows wild in the south of Europe, where it is also largely cultivated. The engraving shows a small branch, with leaves, buds, and a flower, The buds are picked when they are about half grown, by women and children, who find it no pleasant task, on account of the prickles which are found at the base of each leaf. The picking continues throughout a good part of the year, each day's gathering being put into casks and covered with vinegar to which some salt has been added. When the season is over, the capers are sorted into several sizes by means of sieves, and put into fresh vinegar and exported in bottles or small casks.

The plant is half hardy in England and would doubtless succeed in some of our southern states. Capers have a peculiar aromatic taste and have been employed as a pickle for hundreds of years: their chief use at present is to mix with drawn butter to form a sauce for boiled mutton. The fruit of the garden nasturtium *(Tropeolum)* is often used as a substitute, as also is, in England, the fruit of the caper spurge *(Euphorbia lathyris)*. We should doubt, however, about the safety of the last mentioned substitution, as the plant belongs to a family producing many very poisonous plants. (1865)

Chicory

Chicory is botanically known as *Chicorium Intybus*, and is sometimes confounded with the endive, which is a different species, *(Chicorium Endivia.)* The plant is a native of Southern Europe, and has become introduced into this country where, especially near the Eastern cities, it is a very common weed. The fleshy perennial root throws up a stem the second year which bears an abundance of pretty blue flowers, which open only in the sunshine. The general appearance of the plant in the wild state is well represented in the engraving; the detached flower is about half the natural size. Like the dandelion, to which it is closely related, all parts of the plant have a milky juice. In Europe the blanched leaves are used as a salad, but it is for the root that the plant is chiefly cultivated. The roasted root has long been used to mix with coffee, and now that the real article bears so high a price, it is advocated as a substitute.

Fig. 1—CHICORY PLANT—*(Cichorium intybus.)*

The culture is the same as that of carrots, about four pounds of seed being required for an acre. The roots may be taken up in the fall or in the spring before the flower stalk shoots up; some claim that the roots are of better quality when two or three years old. The root is washed, sliced, and dried, and then roasted or burned. In England 1 lb. of lard is added to 50 lbs. of chicory while roasting, in order to improve its appearance. With regard to the propriety of using this as a constant beverage, we have already spoken pretty strongly. It is believed to excite the nerves unduly, to derange the digestive functions, producing headaches and other ills. Some say they have used it with impunity, and that those who are unpleasantly affected by coffee find the change to chicory to be beneficial. Much of the coffee sold ready ground is more or less mixed with chicory, and some prefer it. Those who wish to try it as a substitute for coffee or to mix with it, can cultivate a small patch for the erperiment. (1863)

The Dandelion and its Uses

Most persons look upon the dandelion as a weed to be exterminated rather than as a plant to be cultivated. Though not a native of this country, it has kept pace with civilization, and is to be found almost everywhere. Every meadow and grass plot is studded with its bright yellow blossoms in spring, and those who look upon it as a troublesome weed will have to content themselves with trying to crowd it out by better plants, for unless they can bribe the winds to not blow about the seeds, they have a hopeless task in attempting to exterminate it.

The dandelion is so common a plant that we are accustomed to overlook its beauty; its leaves are not inelegant, while its flowers are quite as pretty as many we cultivate for ornament. Nor is the globular head of ripened fruits the least interesting part of the plant. Each little one-seeded fruit has a delicate little long handled parasol made up of hairs attached to it; a contrivance well adapted to aid in its distribution by the winds. The leaves vary greatly according to the situation in which the plant grows, but they are all marked with strong tooth-like notches which suggested one of the French names of the plant, *Dent de lion* (lion's tooth), from which is derived our word dandelion. The leaves are much used as greens, and when blanched they form a salad not unlike endive. The root is employed medicinally, and is one of the many articles used as substitutes for, or to mix with coffee.

The plant is botanically related to both chicory and endive, and is used in a similar way. Those who value it for greens will find it much better to cultivate the plants than to depend upon those which grow spontaneously, as they are superior, and are always at hand. When the root is required, it should always be taken up in the fall, as then it contains most of the milky juice upon which its properties depend. The seed is sown in May or June, in well prepared ground, in drills 12 or 15 inches apart. Thin to 3 or 4 inches and keep the plants well cultivated through the season, and they will be fit for use in the following spring. According to Burr, if the dandelion is cultivated for its root, the sowing is made in October, the plants thinned the following June, and kept free from weeds during summer, and the roots harvested the next October by plowing them out. The roots are prepared for market by washing, slicing and drying them. (1865)

Mustard

There are two species of mustard raised in the United States; the white *(Sinapis alba)*, which is mostly cultivated as a forage plant; and the black *(S. nigra)*, generally raised for the seed. It requires a rich, loamy soil, deeply plowed, and well harrowed.

It may be sown either broadcast, in drills about two feet apart, or in hills. Mr. Parmelee, of Ohio, thus raised on 27 acres, 23,850 lbs., which brought in the Philadelphia market, $2,908, an average of over $100 per acre. The ground on which it is planted must be frequently stirred, and kept clear of weeds. When matured, it should be carefully cut with scythe or sickle, and if so ripe as to shell, laid into a wagon box with tight canvas over the bottom and sides, so as to prevent waste. As soon as it is perfectly dry, it may be threshed and cleaned, when it is ready for market.

The white mustard is a valuable crop as green food for cattle or sheep, or for plowing in as a fertilizer. For feeding, the white is much preferred to the black, as the seed of the latter is so tenacious of life, as to be eradicated with difficulty when once in the ground. The amount of seed required per acre is from eight to twenty quarts, according to the kind and quality of the land, and the mode of planting or sowing. It may be sown from early spring till August, for the northern and middle states, and till the latter part of September for the southern. The crops yield from 25 to 30 bushels per acre. Both are excellent fertilizers for the soil. (1851)

The Garden Sorrel

The weed too well known to all cultivators as sorrel, has an own brother which is a useful culinary plant: garden sorrel *(Rumex acetosa)*. This has long been cultivated in Continental Europe, and is gradually making its way into our gardens, and even appears, though sparingly, in our markets. The original species, which is a native of Europe, Northern Asia, and British America, has produced in cultivation several varieties, in which the foliage is larger, more succulent, and less intensely acid than in the wild form.

The variety the most esteemed is the Bellville sorrel. The garden sorrel is a perennial, and when one has a few plants to start with, it can be multiplied readily by dividing the old roots in the same manner we do rhubarb. The plants are perfectly hardy, and are not particular to the kind of soil. The leaves, which appear early in the spring, are the portions used, and the larger ones should be cut singly, leaving those in the center to grow. Sorrel is largely used by the French in soups and in salads, and also by itself, cooked in the same manner as spinach. Its tartness is especially relished as an accompaniment to veal. We find that a small quantity of the leaves, cooked with spinach, gives that vegetable an agreeable flavor. The seeds are now kept by our

has an unfailing stream can have an abundant supply of watercress.

In some cases it is merely allowed to grow in the natural stream, but those who make a business of growing it, increase the area by making beds at right angles to the stream. These will depend upon the character of the land and the supply of water. The beds are usually five feet wide, and of a length governed by the level of the land. These beds are excavated to an average depth of about eight inches, and are made about five feet apart. The making of such beds being governed by the peculiarities of each locality, only general directions can be given.

They should be so constructed that they can receive water from the stream, which may be directed into them by the use of board dams. As many beds may be made as can be kept flooded during the winter.

Watercress is naturalized in many streams in the older states, and where it occurs, a supply may be secured for stocking the plantation. The plant is mostly submerged. Each joint below the surface throws off roots, and if the stem be made into cuttings, each of these fragments, if set in the soil of the bed, will soon form a vigorous plant. Such cuttings may be set a foot apart each way in the soil of the beds before the water is let in to them. Those who can not procure cuttings can readily raise the plants from seeds, which are sold by the principal seedsmen. If the seeds are sown in a box in good garden soil, which is kept very moist, a supply of plants for transplanting will soon be at hand. The starting of beds of watercress should begin in early spring. (1883)

Mushroom Culture

Mushrooms are greatly esteemed on account of their peculiar and delicious flavor. They may be stewed, fried in fat, or made into catsup. In some countries, Russia and Poland among the number, there are said

Fresh Parsley in Winter

This can easily be obtained in the following way: Saw a good stout barrel in two, then make auger holes of 3/4ths inch diameter and about 5 inches apart, all over the cask. Before winter sets fairly in, dig up an abundance of plants from the garden, and, beginning at the bottom of the barrel, set the crown of a plant into each hole, covering the roots with good garden soil as you proceed with planting. Go on thus, until the cask is filled, and then set a few plants on the top. Place the cask in a warm and light cellar, or under the staging of a greenhouse, and it will not only look well, but will furnish the table with many a savory garnish all winter. (1860)

dealers, and may be sown in spring the same as beets. (1883)

Watercress—Its Cultivation

In all large cities there is an increasing demand for watercress, and it meets with a ready sale at remunerative prices. As it is a vegetable that can only be grown where there are running streams, its cultivation is limited to comparatively few locations. We know of some farms on the Hudson River, the owners of which receive from their watercress a larger income than they do from all the rest of their crops. Whoever

to be above 30 kinds in use. They are there gathered at different stages of growth, and used raw, boiled, stewed, roasted, and even dried for winter use.

Great care must be exercised in selecting mushrooms for eating, as there are poisonous kinds. Three ways are recommended by which to determine whether they are good: 1., by the color of the gills, that of the good kinds being, when young, of a fine pink, or flesh color, changing however, to that of the questionable kinds — a chocolate color — at more mature growth; 2., by the smell, the good kinds emitting an agreeable odor, while that of the bad is nauseous and disagreeable; and 3., by sprinkling salt upon the inner or spongy part, unwholesome kinds turning yellow, and edible kinds black. Bad kinds are found mostly in forests; edible ones in open pastures, most frequently in old horse pastures, which sometimes, in damp, warm seasons, yield large crops.

Mushrooms may be grown at any season of the year by those who have a suitable bed. This may be made in any dry cellar (or under a shed), the temperature of which can be kept at from 50 to 60 degrees. Extremes of temperature must be avoided. The temperature of the bed should be from 70 to 75 degrees. A bin or bottomless box about 20 inches deep is suitable for a bed, and one 4 feet wide by 8 feet long, if properly managed, will yield a supply for a good-sized family. On the bottom put a thin and slightly oval layer of ashes and gravel, or pieces of bricks, to avoid dampness. Next put in a six-inch layer of unfermented horse droppings from grain and hay-fed horses (not grass-fed), the dryer the better, with short straw intermixed.

When this, being exposed the air but not to rains or dampness, has become quite dry, cover with 2 inches of dry earth — sandy if possible; into the surface of this introduce small pieces of spawn in rows six inches apart. Then repeat the layers of horse droppings and short straw and of earth, with spawn introduced as before. Cover the whole with two or three inches of fresh, warm horse droppings, and occasionally sprinkle with blood-warm water, to induce fermentation, the top layer of fresh manure to be removed as soon as fermentation has caused the spawn to begin to spread. In five or six weeks the mushrooms may be expected. The usual size is from 1½ to 3 inches in diameter, but those over 4 feet in circumference, and weighing 12 to 14 pounds have been produced.

Mushroom spawn is usually to be had in seed stores, and also may be found, of uncertain qualities, however, in dry lumps of horse dung, in old pastures, hot beds and manure heaps. It has the appearance of dry, white threads. (1862)

Make an Asparagus Bed

Asparagus is as easily raised as anything that grows in the garden, and yet it is comparatively rare to find it upon the farmer's table. The reason may be that much nonsense has been published about the difficulties of raising it, and that we have to wait two or three years for the full maturity of the plant. It is true that a full crop will not be given in less than three years, but when the bed is once made, the job is done for a dozen or twenty years.

Any good well-drained soil that will bear corn is suitable for asparagus. Put in a half-cord of manure for every four square rods of ground. Work it in thoroughly. Set out one-year-old plants, in rows four feet apart, and two feet in the row. They can be kept clean then with the harrow or cultivator. It should have cultivation once in two weeks, through the growing season.

Cover the bed with manure in the fall, and fork it under in the spring. Cultivate thoroughly through the second season and top-dress as before. The second season a few stalks may be cut in April and May, but there should be no close cutting until the third year, and this should not be continued later than the middle of June. The plants must have time to grow, and recuperate in mid-summer, or the bed will soon fail. The secret of large, fine asparagus is abundant manure, applied in the fall every season, thorough cultivation until the tops prevent, and stopping the cutting by the middle of June. (1882)

Brussels Sprouts.

Brussels Sprouts

This is one of the several forms in which the cabbage has been developed by cultivation. Instead of forming a single head at the top of the stem, like the ordinary cabbages,

numerous small heads are produced upon the stem for its whole length, as shown in the engraving. These heads are the axilliary buds which develop into little compact cabbages like a miniature Savoy. The leaves of the stem soon fall away, and in a good variety the small heads will be so numerous and close together, as to completely hide the stem.

When or how this variety of the cabbage originated is now known, but it was cultivated in Belgium as early as the year 1213. The cultivation is the same as for late cabbages, and as the plant is quite as hardy as those, it is strange that it is so little cultivated. They may be cooked and served in much the same manner as cabbages, but are much improved in flavor if treated like cauliflower, with a sauce of drawn butter. (1882)

About Savoy Cabbages

An Englishman once wrote a work, "On things not generally known." Were we to write such a work, it should begin with Savoy

CABBAGE. Netted Savoy.

cabbages. Our farmers and others raise cabbages year after year, and very good cabbages of their kind. But the Savoys are of a so much better sort when they come to the table, we are sure that, when once tried, very few would grow any other.

A Savoy cabbage is as much superior to the common hard-headed kinds as the best cultivated grape is superior to the common fox-grape of the woods. Indeed, the English works on gardening treat of cabbages under one head and Savoys under another, as if the two were entirely different vegetables. The Savoys are a distinct race of cabbages; their leaves are always strongly bullated, as the botanists would say, but "blistered" will perhaps describe the peculiarly wrinkled character of the leaves. The heads are never very hard, but the loose outer leaves cook tender. The engraving gives the general appearance of this race of cabbages. (1882)

Henderson's Early Summer Cabbage.

Raising Cauliflowers

Not the least among flowers is the savory cauliflower. So every epicure will say. So every one who enjoys good food will say, and so say we. To grow this vegetable well it is important to give it some care.

Start the seedlings, as we do common cabbages, in a warm rich border. For win-

out in rows three feet apart, and two feet and a half in the row. If the weather be dry, shade them for a few days. Keep the ground loose by hoeing, but do not earth up the plants much, as this often induces rot. Unless the ground is naturally moist, it will be well to cover it in mid-summer with strawy manure, or with clean straw. Once a week, pour over this mulch the refuse water

bottom of a trench, and to cultivate the plants upon level ground. The bed for celery should be well prepared by spading in an abundance of well-decomposed manure. Market gardeners plant the crop upon land that has been heavily manured for early cabbages, cauliflowers, onions, etc. But whether the manure has been applied for a previous crop or not, the soil should be very rich, and if the manure is now applied, it should be well incorporated with the soil.

The plants are set in rows 4 feet apart and 6 inches distant in the rows. The line being stretched, it is to be beaten gently with spade, in order to leave a mark on the soil to serve as a guide in planting. While cabbage and some other plants should be set deeper than they were in the seed bed, this is not the case with celery, which forms no distinct stem, and in transplanting should be set no deeper than it was before. If the transplanting can be done in a moist time, all the better; but if it must be done when the soil is dry, an abundant watering of the plants soon after setting will be of great service. All plants, when transplanted, should have soil well pressed around the roots. Celery, especially, needs attention to this "firming," as gardeners term it. After the celery is planted, there is nothing to be done to the crop for the next two months except to encourage the growth by frequent hoeing, which will remove such weeds as may start, and keep the soil light and open. (1882)

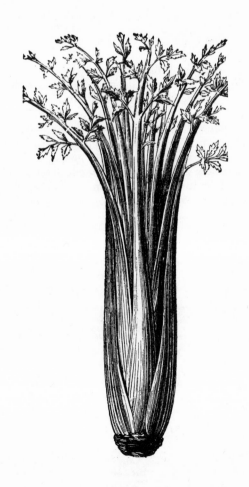

ter and fall use, the plants need not be set out before the 10th of July. Prick them out once or twice in the border, before giving them their final place in the open ground. This will make them strong and stocky plants. In choosing a spot for transplanting them, let it not be in the shade of trees or fences. The soil should be deep and rich. If not so naturally, break it up with a long spade, and put a shovel full of manure at the bottom of each hole, covering the same with two or three inches of fine earth. With a garden trowel, lift the plants and set them

of the kitchen and chamber. In September and October you will enjoy the results of these labors. (1862)

Planting Celery

Probably no gardening operation has been more thoroughly revolutionized within a few years than the growing of celery. The great step in advance in celery culture was to abandon the English method of planting it at the

21

SPECIMENS OF GOURDS AT THE "AMERICAN AGRICULTURIST EXHIBITION."

1. Valparaiso Squash. 2. California Squash. 3. Hubbard Squash. 4. Crook-neck Squash. 5. Turban, or Turk's-head Squash. 6. Golden Winter Scallop. 7. Vegetable Marrow. 8. Green Striped Bush. 9. Lagenaria Vitata. 10. Gourd from Hindostan, new. 11. Mock Orange. 12. Pear Gourd. 13. Sandwich Island Gourd. 14, 15. Unknown. 16. Hercules' Club. 17. Artichoke Gourd. 18. Long Orange Gourd. 19. Cucumis Dipsacius, Japan. 20. Cucurbita Striata. 23. Bottle Gourd. 24. Boston Marrow Squash.

Horse Radish

Perhaps there is no vegetable so really useful, that is treated with so much neglect as the horse radish. Scraped into shreds, or grated fine, and soaked in vinegar, it becomes an excellent condiment for fresh meat and fish; it medicinal uses also, in cases of dropsy, scurvy and rheumatism. It stimulates digestion, excites the glands into action, and warms up the blood in a healthful manner. Aside from all domestic uses, it is worth raising for market. Large quantities are annually bought by pickle manufacturers, ground and packed in closely corked jars and bottles with vinegar, for sale in this country, and for exportation.

It will grow without culture; but to raise it in the best and most profitable manner, it should be treated with some care. Choose a spot of good sandy loam, rather inclining to moisture. Lay off as much space as can be devoted to it, give it a good trenching, and work some old manure down into the lower part of the trench. This will prevent the for-mation of weak sideshoots or prongs, and will favor the growth of large and long roots. After spading and leveling the ground, make holes for the plants eight or ten inches apart, in rows one foot apart. These plants may be cuttings taken from the buds or crowns of old plants; in which case, they should be set in holes near the surface. Or they may be taken from the lower part of the root, in pieces two inches long; in which case, the cuttings should be set a foot or so in depth. Either way will do.

When roots are wanted for table use, uncover one end of a row to the depth of the roots, so that the whole plant may be taken out without breaking. It will be found that plants grown in this way, greatly excel those left to shift for themselves. (1856)

The Gourd Family

Few persons, except professional seedsmen, have an idea of the number of varieties belonging to the gourd tribe, named *Cucurbitaceoe* by botanists. Until recently, comparatively little attention has been paid to their cultivation, except in the case of the squashes and pumpkins, which occupy time-honored places in the garden and the field. Within a few years, fancy and or-namental gourds have been coming into favor for decorative purposes, and their number and beauty have been greatly increased by importation from foreign countries, and hybridization with old varieties.

The above engraving, drawn from specimens at our exhibition at the office of the *Agriculturist*, shows some of the more curious and otherwise noteworthy varieties. Part of these will be recognized as established favorites in the garden and on the table: others are new and striking.

The specimen numbered 5, the turban squash, bears a striking resemblance to a Turkish head dress, and from its beautiful coloring is a most attractive object. It is also edible, and by some considered to be of fine quality. No. 13, the Sandwich Island squash, was trained while growing, into a good resemblance to a swan without wings; the bill is well represented by the stem. No. 10, is a new and singular specimen raised by W. F. Heins from seed sent to our office from Hindostan. From its pungent quality we suspect it belongs rather to the *capsicum* family than to the *cucurbitae*. No. 19 might be called the "vegetable caterpillar." It is about three inches long and half an inch in diameter, of bright green color, and thickly studded with stiff hairy spines. It was grown from seed received from Japan. We have no

22

knowledge of the use made of it there. It forms a unique ornament. No. 18, the long orange gourd with dark green bottom, is one of the most pleasing varieties for ornament. The vine trained upon a rustic trellis or over rock works in some corner of the grounds, is a beautiful object when laden with its rich parti-colored fruit, and the gourds when ripened are very attractive. The markings of green are varied with each specimen, making them still more pleasing. The different kinds of gourds are so easily hybridized, that it is less difficult to procure an almost endless number of sorts, than to preserve any desired variety true to the original. It can only be done by covering the flowers designed for seed, with some protection against insects, and fertilizing them with pollen of their own species. A single bee entering a blossom may bring with him pollen from several different species gathered in other localities, and thus impregnate the flower and cause its seed to vary.

Attractive as the ornamental features of the gourd family, most of our readers will be more particularly interested in edible varieties. For excellence both as a sauce and for pies, the Hubbard squash (No. 3) still remains at the head of the list. It has made its way but slowly into the markets. Its dark green color gives the idea of unripeness, and we have known parties growing it for the first time, to throw away the fruit and pronounce it a humbug, supposing the season to be too short for its maturity. But after having once become acquainted with its excellence, its color is no longer an objection. Next to the Hubbard stands the Boston marrow, already so well known as to need no description. With this, perhaps, even superior to it for pies, the African squash takes rank. It is much larger than the marrow, but this is rather an objection for ordinary family use. One specimen could hardly be wholly used before spoiling.

The cultivation of squashes and pumpkins is not difficult, though a few important particulars must receive attention to secure the best results. Being mostly natives of tropical climates they should have a warm situation, as a southern exposure, or under protection of a building or high wall. It will be very advantageous to start them early in a hot bed or in the house, and transplant them when they have attained the third leaf. Of course, there should be great care to leave the roots entire, and the earth around them undisturbed. An easy way of accomplishing this is to sccop out large turnips, fill them with rich earth, and plant one seed in each. When ready to transplant, cut off the bottom of the turnip, and the roots will soon find their way out; the remaining substance of the turnips will decay and feed the plants.

The best soil for growing these vegetables is a deep, warm, sandy loam, well enriched with stable manure. Not only should the hill be made rich, but also the surrounding soil where the vines will send out rootlets to gather nourishment. Too little room is usually allowed to each plant. They need space enough to run without crowding and shading each other. The area required will of course depend upon the kind cultivated. It is a good plan to sow at intervals a number of extra seeds in each hill as food for insects, which will attack the younger plants, and leave the first to grow too strong to be consumed by them; they can be easily thinned out as needed. (1863)

Kohlrabi

As this is but little known outside of city markets, we give an engraving of it. It is simply a cabbage which does not form a head, but stores up the nutriment that other cabbages place in a large bud at the top of the stem, in the stem itself, which is swollen out to form a globular bulb for the purpose. The stories about its being a cross between a cauliflower and a turnip are all nonsense.

It is only a variety of cabbage that makes a globular stem. As their stem grows rapidly, it is, if taken early enough, tender and very delicious to those who like cabbages, for it is much more like a cabbage than a turnip. When it gets old, it is stringy, and while not fit for the table, is excellent for domestic animals. In England it is, on some soils, grown in preference to turnips, as it is free from the insects that attack the turnip crop.

What a wonderful difference between the kohlrabi, with its stem at least as thick as it is high, and that cabbage which has been grown for centuries on the Isle of Jersey, where the influence of climate and soil has been to produce tall stems. These stems are so long and so useful in various ways, that an English journal some years ago had an article on "Cabbage Timber." The visitor to the Isle sees on every place a patch of these strange cabbages, with stems 6 to 8 feet or more high. The leaves are used for packing butter for market and feeding pigs, while the stems are put to various uses. In the principal town, St. Heliers, the sign, "Maker of Cabbage Canes" is quite common, and few persons visit that charming island without taking away one or more walking sticks made from cabbage stems. Sometimes the stems grow to the height of 10 or 12 feet, when they are used as rafters to sheds, etc. It is said that one of these cabbages had grown to the height of 16 feet, was left for seed, and that a magpie made its nest in the branches which sprung from it in the spring. (1882)

About Leeks

Those who do not like onions will not cultivate leeks, as they have a flavor resembling that of the onion, though quite peculiar. Leeks are so highly prized by the Welsh that they are as much a national vegetable for them as the potato is to the Irish. The leek differs from the onion in having broad flat leaves, and in not swelling out at the bottom. The eatable portion consists of the lower part of the leaves forming a neck which is blanched by earthing up to exclude the light. The engraving shows the appearance of the leek.

Sow seed in early spring in a light rich soil. It may be sown thinly in drills, 15 inches

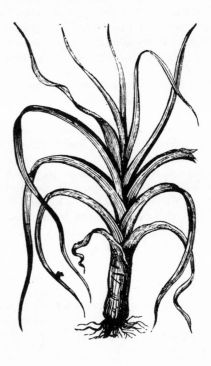

American Flag Leek.

apart, where the plants are to stand; in this case they are thinned out to six inches in the rows, and are gradually earthed up at the summer hoeings. Some cultivators sow the seed broadcast or in drills, and when the plants are four to six inches high, they are transplanted to trenches about six inches deep, and gradually earthed up as they grow. One ounce of seed will produce about 2,000 plants. The leek is quite hardy, and in most localities may be left out over winter, and will come out in spring "as green as a leek." Leeks are used in soups and stews. When cut up in soups and thoroughly cooked, they impart besides their peculiar flavor, a mucilaginous quality much liked by many. (1856)

The Lima Bean

This is probably the best leguminous plant grown in our gardens. It has a rich, sweet, and buttery flavor, which can be compared to nothing but itself. In the neighborhood of each of our large cities several hundred acres are annually devoted to this plant, so great a favorite is it with everybody, in season and out of season.

And it is easily raised. As the summer at the North is a little short for its full enjoyment, it is well to hurry it forward in spring by a shovel or two of manure in each hill, and raising the hill 3 inches above the surrounding surface, will help in giving a start. Let the soil all around be mellow and warm, and the situation be well open to the sun. Plant five or six beans to a hill and if all start, thin them out afterwards to three or four plants. Set the poles in the center of the hill, at the time of planting. Our own practice is to plant in drills on a bed about two feet wide by the side of a cheap trellis or frame, and as the vines grow, they are trained upon twine leaders. This gives them more light and air then when they wind together upon poles, gives stronger growth, and hastens their maturity.

When the first frost comes, gather the pods, and you may have Lima beans at Christmas, as fresh, plump, and delicious, as those picked from the vines in August. How? Pick them before the frost has injured them, spread them on the garret floor, or any airy loft, turning them over once or twice while drying. Reserve the largest and ripest for seed. The green beans will need soaking only 12 hours before cooking, the riper ones will need 24. These beans are particularly fine when used in winter with sweet corn; we have luxuriated upon them recently, and write from inspiration. They may also be kept, when well cooked, in fruit bottles or cans, by corking and sealing them airtight, by which process their peculiar flavor is preserved. With such delicacies within reach, there is no need of being limited to the few articles of diet in common use on the farmer's table. (1862)

Okra, or Gumbo.

Okra—An Excellent Vegetable

This garden vegetable is little known, except by those living near cities, but it is one which most people soon become very fond of when they use it. It is an annual, growing from two to six feet high, with rather coarse leaves, and light yellow flowers having a dark center. The plant belongs to the same family as the hollyhock and cotton, and the flowers of all three bear a strong resemblance.

The young pods are the eatable portion. They are from four to eight inches long, and about an inch in diameter, angled, or several sided and tapering toward the upper end. These when tender are very mucilaginous, and are used for thickening soups and stews. The dish called gumbo in the South consists of chicken stewed with these pods; and the same name is sometimes applied to the plant itself. The pods boiled in water and dressed with drawn butter, after the manner of asparagus, are much liked by many.

Being of southern origin it requires a long season, but lately a dwarf and early variety has been introduced, which is adapted to northern climates. One ounce of seed will sow one hundred feet of row. It should be sown when the ground becomes warm, in rich soil, in drills three feet apart, and the plants should be thinned to one foot in the row. During the summer the plants should be kept clean of weeds and be slightly hilled up in hoeing. The pods are cut when nearly grown, but still tender. The green pods are sliced and dried for winter use. The ripe seeds are among the many things which have been used as substitutes for coffee, and have been advertised as "Illinois Coffee." (1864)

Note on the Cultivation of Onions

Mr. L. T. Keith, of Tompkins Co., N. Y., sends an account of his management of this crop. His onion patch is near his hog pen,

White Globe Onion.

and receives an abundant supply of manure from that source. He manures and plows in the fall, and in the spring gives a thorough harrowing. After raking off the bed it is covered with straw, which is burned over. The seed is sown in rows 18 inches apart, and the bed then receives a dressing of four quarts of ashes, and an equal quantity of hen manure, to every two rods of ground. This application is repeated four times before the onions begin to bottom to any extent. He reports his crop for the last year at three bushels of good onions to every rod of ground. (1856)

Yellow Danvers Onion.

Raising a Crop of Onions

The price of onions is exceedingly variable, and in each season of high prices many have their attention turned to their cultivation. The onion crop is not one that can be profitably grown one year and dropped the next. It is usually the case that those who continue the cultivation year after year, are those who in the long run make it profitable. It is of little use to try to raise onions except on highly manured land, and without being able to give the labor required in weeding just at the needed time. Land that has been for two or more years in corn or potatoes, will answer for the crop. It is claimed by experienced growers that newly-turned sod will not raise good onions.

The land is preferably manured in the fall, using 20 to 30 loads of coarse stable manure to the acre. Or the land is plowed in the fall, and a ton of fish guano to the acre is harrowed in. The land is again plowed shallow in spring, and 300 lbs. to the acre of Peruvian guano, or its equivalent in other good fertilizer, harrowed in. If the ground has not been manured in the fall, then fine pig-pen manure or fine stable manure may be used, plowing in very early, using the guano or other fertilizer afterwards. The harrowing should be very thorough, and if the surface is not smooth, use rakes to finish.

The sowing should be done as early in

Wethersfield Red Onion.

spring as the soil is in good condition. The ground is marked out by a marker in lines 14 inches apart, and the seed sown by one of the several seed sowers; the machine should be set to drop about three seeds to the inch, and they need to be covered about half an inch. All experienced onion growers are very particular about their seeds, which should be new and of home growth. The variety will depend upon the demands of the market.

The cultivation of the crop may be greatly aided by the use of one of the hand weeders or cultivators. If the rows are straight and the sowing regular, a hand cultivator may be run very close to the plants, leaving but a few weeds to be taken out by hand. Usually three or four weedings are needed during the season. Three bushels of salt to the acre, applied when the plants are about four inches high, is beneficial, and at the second weeding it is well to give a good dressing of wood ashes.

We would not advise those who have never raised onions to go largely into their cultivation at first, as they require more attention than many can give, and unless the weeds are kept in subjection the onions will suffer. Some varieties mature much earlier than others; the harvesting is commenced whenever the majority of the tops fall over. (1882)

Plants for Pickles

The universal pickle is the cucumber, and is used when the size of the little finger, up to nearly table size. Our favorite pickle is the Martynia, which should be sown in June, the pods of which, taken early enough, make on the best pickles. Nasturtiums, or Indian Cress, are grown for the purpose. These may be sown in June, and if given some brush to run upon, will give an abundance of their peppery, unripe fruits. Peppers may be set out in June, too; the Squash-peppers with thick flesh, being pickled by themselves, and the large Sweet Mountain kind, used for stuffing.

Besides those plants, grown especially for pickles, many others are used, such as string-beans, cauliflowers, melons, when unripe, onions, cabbages, and green tomatoes. Over-ripe cucumbers are much used for sweet pickles.

An excellent pickle may be made with cabbage, onions, cucumbers, and green tomatoes, about equal parts, a smaller proportion of green peppers, and after chopping and mixing all together, treating as sweet pickle. (1882)

Martynia.

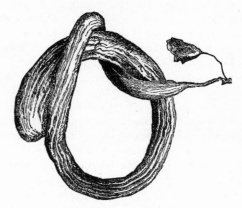

"Snake Cucumber."

A Spring Tart—Rhubarb

Does anybody doubt, or not know the desirableness of this vegetable? Then we pity him. It is one of the finest things in the world, to make a pie or spring tart. Apples often give out in April and May, and those which remain are wilted and tasteless. Man's stomach longs for something fresh, crisp and juicy: the pie-plant affords that very thing. It forms a connecting link in the year-long chain of articles for pie making. Think, too, of the doctors' testimony that it is "one of the most wholesome, cooling and delicious substances that can be used for the table. For dysentery in children, it is an infallible remedy, stewed, seasoned with sugar, and eaten in any quantity with bread." We have tasted samples of fair wine made from this plant. It is also used for jellies and jams.

Procure a few crowns, with roots attached, and set out only one in a place. Rhubarb will live in any kind of soil, but to get large, succulent stalks, the soil must be deep and rich. Five or six plants are enough for an ordinary family. Lay off a bed 20 feet long by 4 wide. Remove the top soil, break up and manure the subsoil heavily, and then return the top spit to its place. This last should be enriched with a light dressing of old manure, and if the land is a stiff clay a little sand should be worked in. Now set out the rows by a line, 4 feet asunder, leaving the plump, pinkish buds an inch or two below the surface. This work may be done in the fall, or early in spring. New roots will soon form, and the growth will rejoice the eyes of the planter.

The after culture is very simple. Keep the ground free from weeds. Pluck no leaves the first year. In the fall put a peck or more of coarse manure around each plant. This will protect the roots, and furnish nutriment for the next year's growth. In the second summer, the leaves may be plucked in moderation, and after that, quite freely. Let the plants, however, have their autumnal dressing, to be forked into the soil the following spring. In our own grounds, we have pursued this course several years; and now the stalks and the leaves of our plants are so magnificent, we are often asked the names of our new and improved varieties. We uniformly reply by simply pointing to the manure heap.

If any one wants to get a very early tart or pie, let him (say, by the middle of March,) set a barrel or rough box, headless and bottomless, over the crowns of several early plants, and surround the same with fresh manure from the horse stable. Put a few forkfuls inside of the barrel, and a bushel or more outside. This will soon generate a local climate of 50 or 60 degrees, and give the plants a start while those out of doors are asleep. The barrels should be kept nearly or quite covered for ten days, and then gradually opened as the season and the plants progress. Add a little fresh manure outside, after the first week. As soon as the outdoor plants are fit to cut, the forced ones should be uncovered and allowed to rest. (1862)

Growing Shallots for Market

With Southern gardeners shallots are a very important crop, and in many cases the most profitable one grown. As they can be shipped at any time between December 1st and March 1st, it is also the earliest crop. The bulbs are put out in rows eighteen inches apart, and six or eight inches apart in the row. They may be planted any time between September 1st and December 1st, although the best season is probably in October.

The ground should be rich as for all garden crops, well drained and in perfect condition; I am not positive as to which kind of soil they grow best in, still, for my part, I should prefer a rich, sandy slope. They are cultivated with hand plows and cultivators, or wheel hoes, though, of course, there is come hand work required that can not be done by machinery.

The white variety sells best and is equal to the red in every other respect. Good stable manure is the best fertilizer that can be used, though many who find it difficult to procure manure, use bone, bat guano, superphosphate of lime, and other fertilizers with success. If stable manure is used, it is generally plowed in; the concentrated fertilizers are sown in the drill. After the bulbs have started they do not require a great deal of work, as they are rapid growers. In cultivating, the earth is gradually worked toward the rows, which causes the onions to bleach higher each time they are worked. This improves the quality, and consequently increases their market value.

Planting and preparing for market are the two great trials of shallot growing, but althoug there is a great deal of light work, there is nothing about the crop that requires hard work. I think it is best to plant the bulbs about two inches deep, but I know on this point there are differences of opinion. One gardener merely opens the drills with a hand plow. He then sets out the bulbs six inches apart, puts in his fertilizer, and waits until they sprout. After they are three or four inches high he draws the earth around them; and he has had as fine shallots as I ever saw. To prepare shallots for market, pull them and shake off all the soil, cut off the roots, and slip off the outer skin. This leaves them perfectly white. They are then tied in bunches of seven or eight, or less if the bulbs are large, and laid in the box with the white end alternating from one side to the other. As the tops are generally too long, they are cut off until they just fit the box. Shallots are shipped in one-third bushel boxes, and they make a very attractive appearance. (1893)

Spinach for Everybody

In the spring every one finds some kind of green vegetable acceptable — it seems to meet a natural want, and a list of the various articles consumed in different parts of the country, under the general name of "greens" would be a long one. A large share of greens is furnished by wild plants, and much time is consumed in hunting these by the roadside and in the fields. A very small portion of this time spent in the garden, would furnish vastly better greens without the trouble of hunting for them.

To have the earliest supply from the garden, the preparation must be made the preceding autumn. But there is no reason why the use of greens should be confined to the first few days of spring. By proper management the garden can be made to yield them the whole season through. Spinach and some others are acceptable at any time, summer or winter, and with a little forethought, may be had except in the coldest months.

It seems strange to those interested in such matters, that certain delicious vegetables, as easily raised as any others, rarely find their way into farmers' gardens. Take spinach for example, one of the most delicate and delicious of all vegetables, always sold, and at a good price, in city markets, is so seldom seen in farm gardens that we may say that it is, as a rule, unknown to them. Yet its culture is as easy as that of its relative, the beet, and it may be had in the greatest abundance at a trifling cost, the seed being cheap.

Select a warm early spot in the garden, and as soon as it can be made ready — which means highly manuring and thorough working — lay out drills 15 inches apart and sow just like beet seed. When the plants are up, stir the soil next to the rows and continue the cultivation by keeping the soil mellow and hand weeding in the rows just as for a crop of beets or carrots. As soon as the leaves are an inch or two long, thin the plants, leaving them about two inches apart, and use the thinnings. Soon the plants in the rows will crowd one another, when every other one may be taken for use, and by the time this thinning is completed, the remaining plants will be ready. In rich soil, the larger the plant the better it will be, and it is in good condition until it begins to show its flower stalks.

A sowing should be made at intervals of two weeks, until hot weather. As to cooking spinach, those who think that greens must be cooked with bacon or pork, will find spinach better cooked thus than any other greens. To do this is to spoil its delicate flavor, and we may add, makes all greens less digestible than when cooked in clear water.

To have spinach in perfection, wash thoroughly, put into boiling water and let it boil with the lid off (to keep it green) 20 or 30 minutes, or until tender. Place on a colander to drain, chop fine, return to the sauce pan — of course having thrown out the water — with a generous lump of butter, and let it simmer until the butter is melted and the whole heated through. It is often served with hard boiled eggs. It may be eaten as other greens, with vinegar, but those who like the delicate flavor of the vegetable do not use any addition. There are other methods of cooking, but this is the simplest, and, to our taste, the best. Let those who have never grown spinach, try it this spring. (1881)

Hubbard Squash.

The Hubbard Squash

Recently not a little interest has been excited in various parts of the country, in reference to the Hubbard squash. Cuts of it, all apparently from duplicates of the same engraving have appeared in many of the agricultural papers; while the uniformity in the descriptions indicates that these also have mainly originated from one source. But though we have heard these facts urged against its claims, the objections are hardly tenable, for we hold that if any man has a good thing, it is his duty, aside from his own interest, to disseminate it as widely as possible by all fair means. It usually or often happens that a really valuable article only becomes known under the stimulus of personal interest. It is the duty of the public, however, to examine into the character of the evidence where it is mainly confined to individual testimony.

Fig. 1.

We were at first disposed to give this new candidate to public favor, the Hubbard squash, a conveniently wide berth, until its claims were generally tested. Recently, however, we have received very strong commendations of it, and we believe they have spoken from personal knowledge.

A friend in Maine sent us a specimen, which he assured us was of the purest quality, and we have prepared the accompanying engravings from it — fig. 1, to illustrate its general outward appearance,

and fig. 2, to show its internal structure, thickness of flesh, etc.

The color is dull, dark green. The skin is thick and hard, as much so as that of a summer squash. It is also closely covered with knots, or warty protuberances not very prominent.

On trial, both by boiling and baking, we will not contest the claim that this variety is superior in quality and flavor, even to our old and long tried favorite, the Boston marrow squash.

Of the origin and history of the Hubbard squash we know nothing beyond what we find in the advertising circular of Mr. Gregory, who last season cultivated it somewhat largely for the seed. He says: "The first specimen was introduced into Marblehead, Mass., over 40 years ago, and its cultivation was probably confined to a single individual for upwards of 20 years. About 15 years since, we received seed from Mrs. Elizabeth Hubbard, (after whom I eventually named it), and have continued its cultivation to this date." (1859)

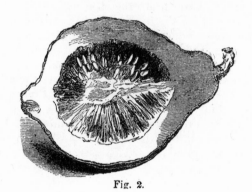

Fig. 2.

Sweet Potatoes

The sweet potato may be cultivated much farther north than is generally supposed. Indeed there is scarcely any locality where a crop of Indian corn can be grown, in which there may not be raised a crop of sweet potatoes, unless the soil is heavy and cold. In such soils success cannot be expected.

If the plants at the start are well rooted, no matter if the sets have lost their leaves, so long as there is a green stem and a good root, they will grow. In all northern localities we advise, by all means, to plant upon ridges. It is a good plan to place rows of well rotted manure three feet apart, and then with the plow turn up ridges over the manure, and finish them with the spade. The ridges being made, level their tops, and set the plants every 15 inches. A common trowel, or an implement made from a shingle or other thin board, is the best thing for setting them. Thrust this straight down, and by moving it back and forth, open a hole; put in the plant down to the last eye. Then put in the trowel, or wooden substitute, a few inches away from

the plant, and crowd the earth firmly up to it, filling the hole thus made with a stroke of the implement. If the soil of the ridges is very dry, it will be well to make the holes for the plants, fill them with water, and let it soak away, and then set the plants. Provide for a few more plants than are likely to be needed; these may be set out, close together in a convenient place, to be drawn upon to replace any that may die in the ridges.

The after-culture consists in keeping the ridges clear of weeds, and after the vines run and cover the ground, occasionally lift them by running a hoe handle or other stick under them, and prevent them from taking root. (1882)

Pruning & Training the Tomato

It is assumed that the plant is trained to a support or trellis of some sort, otherwise pruning would be of little use. One object in pruning should be to remove the superfluous small branches that are produced in abundance, and make a dense, confused mass of foliage. The pruning to remove the excess of fruit may be combined with this; the later flowers are borne upon the small, recently grown branches, and by removing these altogether, two ends are accomplished. When one of the main branches of the plant has set all the fruit it can ripen, it should be stopped, or prevented from growing any longer.

An inspection of the plant will show that it produces its flowers and fruit in a different manner from most others. A flower cluster generally springs from an axil, or where the leaf joins the stem, or, as in the grape, appears at a point opposite to the leaf, while it is in the tomato halfway between two leaves, as

shown in the diagram. In stopping the growth of a stem, there should always be a leaf left above the cluster, the line in the diagram showing where to cut; if desired, as a precaution against accidents, two leaves may be left; in this case removing the upper flower cluster, should there be one above that which has been fixed upon to be the last one upon the stem. When the fruit is partly grown, it will be well to remove all that are deformed and misshapen, and if a cluster promises to be unusually heavy, it will be well to support it to the trellis by a bit of string.

Thinning the crowded branches, removing the excess of fruit, and stopping the growth of stems after they have set sufficient fruit, are the ends to be aimed at in pruning the tomato. In applying these general rules, which are all that can be given, one, in the first attempt, is not in danger of pruning to excess, but of leaving too much. (1882)

Training Tomatoes

Tomato plants are too frequently left to take care of themselves. Some gardeners even maintain that it is better to let them lie upon the ground to receive the benefit of the heat absorbed by the soil during the day. There is some advantage to this, but it is more than counterbalanced by their liability to rot, and to be injured by insects and absence of sunshine. Many who attempt to train their tomatoes, do it imperfectly. A single stake is driven in the ground, and the whole plant is drawn together with a string, and tied to the stake, leaving but a small portion exposed to the sun and air.

A better method, which we have seen practiced, is to "stick" them like peas, using two

or three strong bushes for each plant. In the drawing the plant is spread out upon a frame or low trellis, exposing both fruit and foliage to the full benefit of the sun. Of course the frame should be upon the north side, and it is better to cut away some of the branches, and head back those which are too rampant. Trained and pinched back in this way, ripe fruit can be had, from hot-house plants, during the latter part of July. Another method, though not equal to the above, is to drive stakes about the patch, and fasten piles upon them, making a horizontal frame about two feet from the ground. This keeps them up, but they do not receive the exposure needed, so well as when trained to the upright frame in the manner above described. (1860)

Truffles

Truffles have been the favorite dish of epicures from the beginning of the world to the present day; and yet, strange to say, they are always scarce and high priced, for few know how to raise them, and still fewer have the proper knowledge to prepare them for the table. The royal cooks of France say, that "the truffle improves all it touches;" and happy is the *cuisinier* who can give a taste of its delicacy and flavor to each separate dish which issues from his scientific laboratory. Even the grave and satirical Juvenal writes: *"Tibi habe frumentum, Alledius inquit; O Libye! Disjunge boves, dum tubera mittis."* This we take the liberty of thus freely translating, not doubting that it is the veritable English the gastronomical satirist would have used in our day.
Keep back thy wheat, oh, Egypt! and thy corn, And thy fat beeves which low at early morn; But give me truffles, or I die forlorn!

A gay French writer says: *"Quand je mange des truffes je me crois transporté dans un autre monde,"* &c.; but we forget — this is not English, and therefore we translate the lively Gaul for the benefit of our readers as freely as we have the grave old Roman. "When I eat truffles I at once think myself transported to another world." (1845)

The Root Harvest

A large part of the work of harvesting of the root crop is done in November. Potatoes are not roots, but are often, and quite naturally, classed with them. They should be dug so soon as growth is completed, and ought to be in winter quarters before the middle of the

month, and in most sections much earlier than this. The rot is favored by the tubers remaining in the soil, especially if the weather is wet and the ground is much soaked by water.

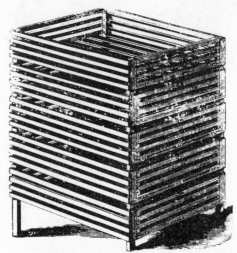

Fig. 1.—A SERIES OF POTATO BINS.

Store the potatoes in bins, fig. 1, in a dry, cool cellar, with ample ventilation. They may be put in pits in a dry place in the field, and covered with straw, boards, and enough earth to keep the tubers from freezing. If there are rotten potatoes at digging time, they should be thrown in the burn heap and destroyed, to prevent the spread of the rot fungus. Those that are only slightly affected should also be removed and fed to farm stock. If a decaying potato remains in the bin or pit, it will communicate the rot to others and may do much damage.

Beets and mangels should be harvested as soon as the growth is finished, or the leaves have wilted from the action of frost. These roots will bear only slight frost without injury. Much of the labor of digging may be done by a horse and plow. Run a furrow with a common plow close to one side of a row, and afterwards pass a subsoil plow near the other side when the roots are loosened and can be quickly removed from the soil.

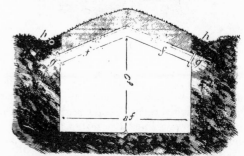

Fig. 2.—CROSS-SECTION OF ROOT CELLAR.

A substantial structure for keeping roots and other like products is shown in figs. 2 and 3. Fig. 2 shows a cross section of a field root cellar dug in dry ground. The excavation is 5 feet deep, 8 feet wide, and 10 feet longer than it is desired to have the cellar. One-half foot below the surface cut a place, *g,g,* to form an oblique support for boards 8 inches wide. Rafters, *f,f,* 5 feet long, are set in pairs upon these boards every 3 feet the whole length of the cellar. The roof is covered 12 to 18 inches with earth, and sodded, with a gutter provided at each side, *h,h,* for the removal of water. The ends are closed with double boarding, filled with sawdust, and provided with doors. Openings over the doors, supplied with movable shutters, give the necessary ventilation. A longitudinal section of the root cellar is shown in fig. 3; *a* is the approach; *b,* steps; *c,* body of the cellar; *d,* the board roof; *e,* the earth covering; *f,* rafters. In light sandy soils it will be necessary to wall up the sides with stone or posts and boards. Such a cellar will last for many years, and be thoroughly frost-proof. If 30 feet long, it will hold about 700 bushels of roots. (1882)

Fig. 3.—LONGITUDINAL SECTION OF ROOT CELLAR.

BROCCOLI.

German, *Brocoli, Sparget-Kohl.*—French, *Chou Brocoli.*—Spanish, *Broculi.*

Nearly allied to the Cauliflower, but more hardy; the seed should be sown in this district in the early part of May, and transplanted in June ; further South the sowing should be delayed until June or July, and the transplanting accordingly, from August to October. In parts of the country where the thermometer does not fall below 20 or 25 degrees, Broccoli may be had in perfection from November until March. It succeeds best in a moist and rather cool atmosphere.

If by mail in quantities of ¼ lb. and upwards, postage must be added at the rate of 16c. per lb.

WHITE CAPE. Heads medium size, close, compact, and of creamy white color; one of the most certain to head. Pkt., 15c.; oz., 75c.; ¼ lb., $2.50.

PURPLE CAPE. Differs only in color. Pkt., 10c.; oz., 50c.; ¼ lb., $1.75.

Walcheren. A valuable variety, with very large, firm heads. Pkt., 10c.; oz., 60c.: ¼ lb., $2.00.

Celeriac, or Turnip Rooted Celery.

CELERIAC, TURNIP-ROOTED CELERY.

French, *Céleri-rave*—German, *Knol-Seleri.*

If by mail in quantities of ¼ lb. and upwards, postage must be added at the rate of 16c. per lb.

A variety of Celery having turnip-shaped roots, which may be cooked and sliced, and used with vinegar, making a most excellent salad. It is more hardy and may be treated in the same manner as Celery. Pkt., 10c.; oz., 25c.; ¼ lb., 75c.; lb., $2.50.

New Apple Shaped. A great improvement over the old variety, having small foliage, large tubers almost round in shape, and smooth. Pkt., 10c.; oz., 30c.; ¼ lb., $1.00; lb., $3.00

ENDIVE.

German, *Endivien.* French, *Chicorée.*—Spanish, *Endivia.*

Endive is one of the best salads for fall and winter use. Sow for an early supply about the middle of April. As it is used mostly in the fall months, the main sowings are made in June and July, from which plantations are formed at one foot apart each way, in August and September. It requires no special soil or manure, and after planting is kept clear of weeds until the plant has attained its full size, when the process of blanching begins. This is effected by gathering up the leaves and tying them by their tips in a conical form, with bass matting. This excludes the light and air from the inner leaves. which, in the course of from three to six weeks, according to the temperature at the time, become blanched. Another and simpler method consists in covering up the plants, as they grow, with slats or boards, which serve the same purpose, by excluding the light, as the tying up.

If by mail in quantities of ¼ lb. and upwards, postage must be added at the rate of 16c. per lb.

GREEN CURLED. Very hardy; leaves dark green, tender and crisp. Pkt., 5c.; oz., 20c.; ¼ lb., 60c.; lb., $2.00.

White Curled. Leaves pale green; should be used when young. Pkt., 10c.; oz., 35c.; ¼ lb., $1.00; lb., $3.50.

FRENCH MOSS CURLED. A beautiful variety of fine quality. Pkt., 10c.; oz., 30c.; ¼ lb., $1.00; lb., $3.00.

BROAD LEAVED BATAVIAN. (*Escarolle.*) Chiefly used in soups and stews; requires to be tied up for blanching. Pkt. 10c.; oz., 30c.; ¼ lb., $1.00; lb., $3.00.

GARLIC.

German, *Knoblauch.*—French, *Ail.*—Spanish, *Ajo.*

Used for flavoring soups, stews and other dishes. Garlic thrives best in a light, well-enriched soil; the sets should be planted in early spring, in rows one foot apart, and from one to five inches between the plants in the rows. The crop matures in August, when it is harvested like the Onion.

If by mail in quantities of ¼ lb. or upwards, postage must be added at the rate of 16c. per lb.

Garlic Sets. Per lb., 40c.

SALSIFY, or Oyster Plant.

German, *Borsbart.*—French, *Salsifis.*—Spanish, *Ostra Vegetal.*

The Oyster Plant succeeds best in light, well enriched, mellow soil, which, previous to sowing the seeds, should be stirred to a depth of eighteen inches. Sow early in spring, in drills fifteen inches apart ; cover the seeds with fine soil, an inch and a half in depth, and when the plants are strong enough, thin out to six inches apart. (*See Cut.*).

If by mail in quantities of ¼ lb. and upwards, postage must be added at the rate of 16c. per lb.

Pkt., 10c.; oz., 30c.; ¼ lb., $1.00; lb., $3.50.

SWEET, POT and MEDICINAL HERBS.

No garden is complete without a few herbs for culinary or medicinal purposes; and care should be taken to harvest them properly. This should be done on a dry day, just before they come in full bloom, then dried quickly and packed closely, entirely excluded from the air. Sow in spring, in shallow drills, one foot apart, and when well up, thin out or transplant to a proper distance apart.

Anise (*Pimpenellum Anisum*), cultivated principally for garnishing and for seasoning, like Fennel. Pkt., 5c.; oz., 15c.

Balm (*Melissa Officinalis*), principally used for making balm tea or balm wine. Pkt., 10c.; oz., 50c.

Basil, Sweet (*Ocymum Basilicum.*) The leaves and tops of the shoots are the parts gathered, and are used for highly seasoned dishes, as well as in soups, stews, and sauces; a leaf or two is sometimes introduced into salads. Pkt., 10c.; oz., 50c.

Bene (*Sesamum Orientale*). Pkt., 5c.; oz., 20c.

Borage (*Borago Officinalis*). Excellent for bees. Pkt., 5c.; oz., 25c.

Caraway (*Carum Carai*). Chiefly cultivated for the seed, which is used in confectionery and medicine; in spring the under leaves are sometimes put in soups. Pkt., 5c.; oz., 15c.

Castor Oil Plant (*Ricinus Communis*). Pkt., 5c.; oz., 15c.

Catnip (*Nepeta Cataria*). Pkt., 15c.; oz., 60c.

Coriander (*Coriandrum Sativum*). Cultivated for garnishing, but more frequently for its seeds, which are used by confectioners. Pkt., 5c.; oz., 15c.

Dill (*Anethum Graveolens*). The leaves are used in soups and sauces, and to put along with pickles. Pkt., 5c.; oz., 15c.

Fennel (*Anethum Fœniculum*). The leaves, boiled, enter into many fish sauces, and raw, form a beautiful ornament. Pkt., 5c.; oz., 15c.

Hop Seed (*Humulus Nupulus*). Pkt., 25c.; oz., $2.00.

Horehound (*Marrubium Vulgare*). Principally used for medicinal purposes. Pkt., 10c.; oz., 50c.

Hyssop (*Hyssopus Officinalis*). The leafy tops and flowers are gathered and dried for making Hyssop Tea and other purposes, Pkt., 10c.; oz., 40c.

Lavender (*Lavandula Spica*). A popular aromatic herb. Pkt., 10c.; oz., 40c.

Marjoram, Sweet (*Origanum Majorano*). For seasoning. Pkt., 10c.; oz., 40c.

—— **Pot** (*Origanum Onites*). Pkt., 10c.; oz., 50.

Opium Poppy (*Papaver Somniferum*). Pkt., 5c.; oz., 30c.

Rosemary (*Rosmarinus Officinalis*). An aromatic herb. Pkt., 10c.; oz., 50c.

Rue (*Ruta Graveolens*). Used for medical purposes, also given to fowl for the croup, Pkt., 10c.; oz., 40c.

Saffron (*Carthamus Tinctorius*). Pkt., 5c.; oz., 15c.

Sage (*Salvia Officinalis*). The leaves and tender tops are used in stuffing and sauces. Pkt., 5c.; oz., 30c.; lb., $3.00.

Savory, Summer (*Satueria Hortensis*). Used for seasoning. Pkt., 5c.; oz., 25c.

Sorrel, Broad Leaved. Used for salads. Pkt., 5c.; oz., 15c.

Thyme, Broad-Leaved (*Thymus Vulgaris*). For seasoning, etc. Pkt., 10c., oz., 40c.; lb., $1.00.

Tansy (*Tanacetum Vulgaris*). Pkt., 10c.; oz., 50c.

Tarragon. Used in salads, soups, etc. Pkt., 25c.

Winter Savory. Pkt., 10c.

Wormwood (*Artemesia Absinthium*). Used for medical purposes; it is also beneficial to poultry, and should be planted in poultry grounds. Pkt., 10c.; oz., 50c.

II Flower Gardening

SPRING.

Keeping the Flower Garden

A small garden with few well-chosen, well-grown flowering plants, if kept in scrupulous order and neatness, will be far more gratifying to its owner, and to his friends, than a large collection of plants without order, arrangement or neatness. Of course, there is a choice in plants, and a garden of the common flowers, well kept, is far more creditable than a jungle of expensive novelties, left to care for themselves. "Good keeping" in the garden is equivalent to "good keeping" in the house. Both depend upon hundreds of small things, easily done if attended to at the right time.

In a large garden a special tool house will be found most useful, but in a small one a place for the tools must be found in some other building. The wall of a woodshed or of a barn, or a suitable, large, cupboard-like structure, built against a board fence or a building, may be made — indeed, any place

that will hold the tools. It is far more important that the tools are brought back to their places after using, than that their places should be showy.

In every garden, even a very small one, there occur various materials, collectively known as "rubbish," which must be disposed of. It is a mistake to have a rubbish heap in some corner, and throw all the refuse matter from the garden upon it. There should be at least three places of deposit for the different kinds of rubbish, if the premises are large enough, otherwise it must be removed altogether. Usually the refuse from the vegetable garden, and that from the flower garden can be brought together. There should be a place at a safe distance from buildings and fences, for the "burn heap." Prunings of trees and shrubs, old and useless labels, stakes and supports of all kinds should go there. There is a vast amount of rubbish, the best use to make of which, is to convert it into ashes, of which one can hardly

have too much. By giving attention to accumulating for this heap, and burning at convenient times, it is astonishing what an amount of ashes may be saved. Keep the ashes in a dry place, but not in a wooden receptacle in a wooden building.

In both the flower and vegetable garden, a large quantity of material is thrown out, which is vegetable matter, and if properly treated will decay, and form a valuable plant food. There should be a trench dug for the compost heap, in which place a few inches of coarse and fresh manure. To this are to be added all the trimmings of vegetables, weeds that have not gone to seed, fallen leaves, indeed any vegetable matter that can not be fed to animals, encouraging fermentation with an occasional layer of fresh stable manure. In a large garden there will be material enough for more than one such heap.

Upon every place there will accumulate a lot of indestructible rubbish, which will neither decay nor burn. The best way to dispose of this, is to dig a moderately deep well, and provide it with a safe cover. All the stones from the garden, broken crockery and tin-cans from the house, and other such waste, should go here. When the deposit reaches within two or three feet of the top, the well may be filled up, and another made, thus allowing of the disposal a vast amount of rubbish, which otherwise will become a nuisance. (1887)

Plan of Flower Garden

In arranging the flower garden, it is an excellent plan to set "bedding-plants" in circular and other fancy-shaped beds cut out in the lawn. Herbaceous perennials, biennials, annuals, small shrubs and vines, are best kept in a flower garden by themselves, where their beauties may wax and wane without disturbing anybody's fastidious taste.

This little garden should be laid off on one side of the pleasure grounds, and be partially hid from the highly dressed lawn by thickets of shrubbery. Whatever is disagreeable in its dug beds, and perhaps straggling and sometimes decaying plants, will then be concealed from the chance visitor to the house; and whatever is agreeable in a surprise, and a cosy corner containing all sorts of flowers, will then be realized.

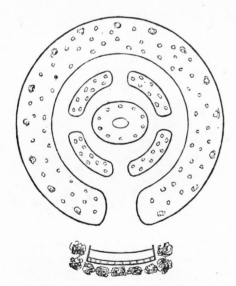

After trying in his own grounds, square, diamond-shaped, oval and other sorts of beds, the writer hit upon the above plan which suits well. The center bed (oval) has a collection of blue and white flowering plants such as larkspurs, pyrethrums, and achilleas. In the middle is a clematis *flammula*, trained upon a harp-shaped frame about seven feet high. This bed is seven feet long and five wide. The four beds, next around it, are occupied with miscellaneous plants; and as they are mostly such as we can recommend for similar gardens, we will mention the

names of the larger portion. One of these four beds is entirely filled with phloxes, early and late. Another bed has herbaceous spiraeas. Another has campanulas, lily of the valley, ragged robin, scarlet lychnis, monkshood and fraxinellas. Another has potentillas, ranunculus, lythrum, forget-me-not, sweet William, baptiseas, sweet violets, columbines, etc. These beds are about ten feet long and four and a half wide.

The outer bed is seven feet in width. Its inner border is decked with low, early flowering plants, such as daisies, polyanthus, iris, daffodils and snow-drops. Next behind these are set various medium-sized herbaceous plants, with spaces left between them for annual flowers. In the rear of all, are shrubs and vines such as roses, spiraeas, deutzias, Weigelias, laburnums, forsythia, flowering almond, Japan quince, honeysuckles, etc. This flower garden occupies a corner of our grounds, and is so completely surrounded by shrubbery that it can not be seen from the neighboring street, so that ladies and children may enjoy it at any hour of the day, without being exposed to the gaze of the public highway. On one side, under a group of low trees and shrubs, is a rustic seat, where a company can sit in the shade, and enjoy a view of the garden in the hottest day of summer. (1859)

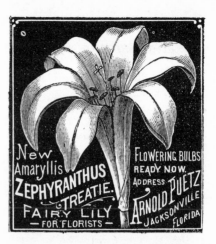

LILY OF THE VALLEY

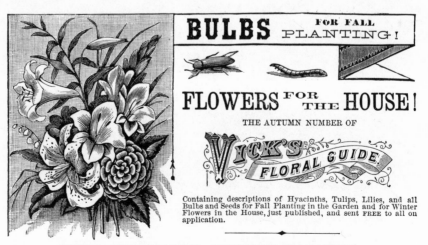

French and German Asters

The engraving presented herewith exhibits, so far as can be done in black and white, the appearance of a beautiful bouquet of aster flowers shown at the *Agriculturist* office recently, by John Wesley Jones, of Columbia Co., N.Y.

One can hardly have an idea of the real beauty of these flowers, without seeing the blooms themselves, or a colored picture of them, showing the brilliant crimson, purple, blue, etc., with the intermediate colors.

For the purposes of raising seed for distribution to our subscribers, we sowed nearly ¼ acre with these improved asters, and never have we seen anything more beautiful in the floral line. Perhaps no flower has undergone more changes and improvements within a brief period than the aster.

These great changes have been brought about by carefully selecting seeds of the finest blooms, and giving them high culture, so as to change the stamens into petals. French and German florists, conspicuous among whom is M. Truffaut, have brought about these improvements and made this one of the most desirable flowers.

Besides being one of the prettiest annuals, the aster is of very easy culture. It will flourish in any good garden soil, and may be sown in the open ground at any time in May or June even, covering with ¼-inch of finely pulverized soil. They are best sown in drills, 15 inches apart, or thinly massed in small plots, with the tallest sorts in the center.

When practicable, it is desirable to sow a portion of the seeds in pots, under glass, early in April, or the last of March, and then transplant them into the open ground, in the border and elsewhere, when the ground is warm, and the weather settled. These, with seed sown in the open ground in May and June, will furnish a succession of beautiful blooms all through the summer and autumn, until severe frost cuts them down. It is unfortunate that the finest blooms supply so little seed for wide diffusion. The petals or flower leaves in the most double sorts nearly or quite fill up the center, leaving little space for seed. (1862)

Nos. 1 and 5, Chrysanthemum Asters.
Nos. 3 and 4, Globe Quilted Asters.

Nos. 6 and 9, Bouquet Pyramidal Asters.
Nos. 2, 7, 9, and 10, Truffaut Pæony Asters.

Chrysanthemums

During this winter, so far, we have greatly enjoyed the flowers of these plants, which have survived the heavy frosts of autumn, and even now challenge the snows to deprive them of their freshness, and we feel moved to speak of them to our readers. The old fashioned sorts, (the large flowering or Indian varieties) are still desirable in a collection, although many cultivators now prefer the Pom-pones or Chinese, especially for house and pot culture. We present above a beautiful engraving of the Vespa, one of the finest of the large flowering sorts. It is a free blooming pure white variety, scarcely known in this country.

What then of culture? About the middle of May, in Northern latitudes, take cuttings from the old roots, and set them in the open border. If the soil has a plentiful admixture of sand, they will strike more freely. Keep the ground moist around them by a mulching of cut grass, and an occasional watering in dry weather. They will form roots and begin to grow in a few weeks. After they have grown two or three inches, pinch out the top, so as to cause the lower branches to break and to give the plants a broad base; for nothing looks worse in a chrysanthemum than a lean and naked stem. If they have grown three or four inches more, pinch again; but do not pursue this operation after August. After that time they should be allowed to form their flower buds.

About July commence potting them, using a good loam mixed with one-third rotten dung. If you have time and patience to give them the best possible culture, begin with very small pots and shift them several times until the first of September, when they should receive their final shift in quart pots, or a size larger. If you have not patience, defer potting until August, and then give them quart pots at once.

After the middle of September, they should be taken into the house, or removed to some warm and sheltered spot, safe from frost. If they can be kept out until October, it will be all the better. If you want very fine foliage and flowers, give the plants a weekly application of manure-water, beginning in August or first of September, and continue until the blooming period is passed. After they have done flowering — which with most varieties will be about New Year's — they may be set under the staging of the greenhouse, or carried into the cellar, to remain dormant until cuttings are wanted in the spring. (1859)

The Dahlia

As May is the season above all others of flowers, we have thought we could not do a greater favor to our fair friends, the ladies, than to introduce to their notice the dahlia; which if it were fragrant like the rose, would at once become the queen of flowers. It was

DOUBLE DAHLIA.

first discovered in 1789, by Baron Humboldt, then travelling in Mexico, and sent by him to Europe, whence it has been disseminated over the whole civilized world.

It is of large size and of every variety of color, from the pure white up to the deepest purple. It may be propagated either by seeds or roots, and some assert successfully by cuttings from the lower part of the stem. We have only tried the two former methods of growing it, and in these we have been quite successful. It is a hardy plant, and has done best with us on a soil of rich clay loam, and during those seasons that moisture and cool weather most predominate. We found the dahlia usually of a larger growth in England than in our own country; but whether owing to the superiority of the cultivation there, or the greater humidity and coolness of the summers, we were unable to decide. We suspect, however, that both of these causes had their effect in adding to the size of this magnificent flower. (1843)

The Hollyhock

Does anybody fully appreciate this flower? The poetical and domestic associations connected with it are part of its recommendations. It has long been cultivated in the gardens of our fathers. The poets have loved it and sung its praises. It was Wordsworth's pet among the flowers; he had groups and rows of them in his garden, where he walked with great delight. And then, it is a robust flower, asking no tender nursing, and blooming profusely without any care.

The commonest and simplest way of propagating the hollyhock is by division of the roots. But as this would not satisfy the yearly demand for the choice varieties, it is propagated extensively by cuttings. These are made in the summer, as soon as the blossoming season is over. The cuttings, each two inches long and containing a single bud, are split in two, the pith taken out, and then they are placed in propagating pans of sandy soil about an inch and a half deep, and covered with a hand-glass. In a few weeks roots are formed, and the plants are potted and ready for sale. The hollyhock will live and grow almost anywhere, but to attain perfection, it should be planted in a deep, strong soil well manured and worked. Good treatment of any plant is always well repaid. (1859)

The Honeysuckle

This beautiful flower is so well known that it needs no description. We have found 19 varieties cultivated in the gardens in this vicinity, and how many more exist we are unable to say. It also grows wild the United States over, and is a great favorite, as well it may be, with the ladies. We wish it were oftener transplanted to the yards and gardens.

In remote settlement where cultivated flowers are scarce, resort should always be made to the wild ones around. Transplanting and good cultivation frequently greatly improve them, and under any circumstances, many which are indigenous to our country, are superior to the exotics for which we pay high prices abroad, and in addition incur considerable risk and expense in their transportation home. In cultivating wild flowers all this is saved, and a benefit in addition is conferred upon the flora of America. Neatly whitewashed, and surrounded by native flowers, even the humble log house becomes a pleasing feature in the landscape, and adds much to its picturesque variety. (1841)

Lavender—Its Cultivation

In this country, the nearest approach to the "Physic Gardens" of Europe is to be found in the limited areas devoted to medicinal plants by the Shaker communities. There is nothing in our climate or soils to prevent the growth of the medicinal plants so largely produced in Europe. But aside from their cultivation, the gathering of the products, and their preparation for market, demand an amount of manual labor that prevents their profitable culture in this country.

We have had, of late, numerous inquiries as to the cultivation of lavender, which we can best answer in a general article. Lavender *(Lavandula vera)* is a low shrub, about three feet high, a native of Southern Europe. The engraving gives the general aspect of the plant; the narrow leaves are of a hoary green color, and the flowers, borne upon long and slender spikes, are blue. This shade of blue is so peculiar that "lavender-blue," or "lavender-color," is in common use to describe this tint.

The flowers are delightfully fragrant, owing to a volatile oil, which, in various forms, is universally popular as a perfume, and on account of its stimulant, aromatic qualities, is used in medicine. The common name of the plant, and the botanical name *(Lavandula),* come from the Latin, *lavare,* to wash, the perfume having long been in use in bathing, and the use of flowers for perfuming newly-washed linen, has given origin to the expression, "laid up in lavender."

THE LAVENDER PLANT.

Lavender may be raised from seeds, sown in a seed bed in spring, and the plants set when two inches high, in rows two feet apart, and a foot distant in the rows. When the plants crowd one another, every alternate one is taken up and transplanted to new rows, placing them two feet apart. After a plantation is established, the plants may be increased by means of cuttings, which take root very readily. A light, warm soil is better than a heavy one. The crop is gathered by cutting

the flower spikes, just as the bloom commences, by means of a sickle, and tying them in bunches of convenient size. In England, as a general thing, the producers of the flowers sell them to those who distil the oil from them. It is said that from twelve to twenty-four pounds of oil are produced to the acre.

Though a native of a warm climate, lavender is not especially tender; we have known it to endure the somewhat severe winters of Newburgh, N.Y., and it will no doubt succeed in any of the Middle States. Lavender has long been cultivated to a moderate extent near Philadelphia, and in their season the flowers are offered for sale in the markets of that city. The flowers are purchased by those who wish to make sachets for perfuming clothing, and by the druggists who use them for distilling lavender water. It is estimated that within 20 miles of London, there are about 600 acres devoted to the cultivation of lavender. The oil of lavender produced in England brings a much higher price than that from the continent. (1883)

A Beautiful Narcissus

The genus Narcissus, in its several species and hybrids, offers a great variety of beautiful flowers, and is worthy of special attention. Some of them make charming plants for forcing. What for example can be more pleasing as a green house decoration, than the old, though by no means common, Hooppetticoat narcissus *(Narcissus bulbocodium)?* A pot containing half a dozen of these little bulbs, makes a bright spot in the greenhouse, and one that retains its beauty for a long time. It is also excellent as a window plant; the bulbs are to be potted and kept in the cellar, until they form abundant roots; they may then be brought into the warmest and lightest place in the window, and will soon show their bright and delicate flowers. (1882)

Double-Flowered Petunias

Most persons are familiar with single-flowered petunias. Like the verbena they produce beautiful flowers during the entire summer. These are white, purple and crimson, with all the intermediate shades. They can be raised from the seed, but do not thus grow true to the parent plant, and layers or cuttings must be resorted to in order to propagate any choice variety.

The best time for increasing them for summer plants is early in spring when slips of young half-hardened wood are to be set in pots supplied with equal parts of loam and white sand well mixed, first filling the pots half full of potsherds or coarse gravel to preserve a good drainage. A bell or other glass must be placed over the cuttings, and care be taken in watering and ventilation to prevent their damping off. In the fore part of May, or as soon as all danger of frost is past, transplant in the open ground and they will soon commence flowering.

The plants are tender and require winter protection. In autumn any choice plant desired for propagation should be taken up, partly trimmed down, potted, and kept in a greenhouse or warm room until spring, by which time a new crop of shoots will be produced for putting out. They will grow well in almost any soil. Last year, for the first time we believe, a new variety — a double-flowered petunia — was produced in France called the *Imperialis*. One of these was taken to England, and a batch of seedlings raised by Mr. Grieves, the first season, by hybridizing the Imperialis with the best varieties of single-flowered, the result being a number of beautiful varieties, distinct in form and color, the color going through all the shades of crimson, purple and white. The habit of the plant is very robust, vigorous, and much more compact than the single-flowered. It appears to grow and flower abundantly in ordinary garden soil. (1857)

Pinks—Carnations & Picotees

Few real lovers of flowers, not those guided by fashion, who, if asked to name their special choice among florists' flowers, would not place the pinks very high upon the list of their favorites. The genus *Dianthus*, the genus of the pink, is a large one, species being found in every quarter of the globe. Botanists have, first and last, described over two hundred species; a careful revision, within the last quarter century, has reduced this number to about seventy, of which many are in cultivation.

Indeed, the pinks known as carnations, etc., are among the oldest of florists' flowers, by which we mean those flowers, the varieties of which are kept distinct, and multiplied varieties are almost countless, and perhaps no one country produces a greater number, naturally, than our own; some of which, especially those from the prairies, have been introduced into England and other parts of Europe, and are highly valued there.

Every lady should have them in her yard and garden in summer, and in her parlor in winter. They are delightful to the eye, and fragrant to the smell, and add cheerfulness and beauty to all around them. Then cultivate the rose—the "lady rose" above all others. (1843)

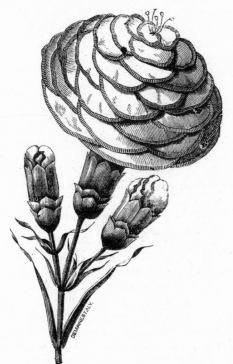

TYPE OF LACED "PAISLEY" OR FLORIST PINK.

and improved by cultivation. It is the varieties known as carnations, etc., that we have under consideration; those in their present great variety are supposed to have originated from the old "Clove Pink," *Dianthus caryophyllus*, a native of Great Britain and Continental Europe. Its stems are nearly woody below, and its pale, grass-like leaves are seen in its cultivated descendants. A strong plant will be from two to three feet high, its erect branches, each terminated by a flower of great beauty of color, and of most charming fragrance.

The new "Perpetual Carnations," will be found described in the catalogues, and some are of great promise. All are fragrant, the picotees being less so than others. It is to their fragrance that the flowers owe so much of their popularity.

Pinks may be readily raised from seeds even in August. The young plants should be grown until just before the frost destroys them, when the plants may be transferred to a frame; or place a frame over the seed bed, and fill the frame with leaves, put on the sash and leave all until spring. Late in April transplant the young pinks to the bed where they are to bloom. Promising plants may be selected to propagate from by layers or by cuttings. (1888)

The Rose

We here present to our fair readers the queen of flowers, the rose—the "lady rose," the truest emblem of themselves and charms. It is a flower so well known, and so universally cultivated, that it is quite unnecessary for us to say one word upon the subject. The

NEW ROSE, "Perle des Jardins."

38

ANTIRRHINUM GRÆCUM—SNAP-DRAGON.

ANTIRRHINUM.

Snap-Dragon

In the *Agriculturist* seed list is the snap-dragon, *(Antirrhinum majus)*, so called from a peculiarity of its flower, which, when gently pressed with the thumb and finger opens and shuts somewhat like an animal's mouth. The toad flax *(Antirrhinum linaria)* of the road side, with its bright yellow and orange flowers, is a good type of this family, though greatly inferior to the cultivated varieties.

The garden snap-dragon is considered a perennial, but is very apt to die out after a few years. It often flowers the same season it is sown. Latterly some very fine varieties have been produced, of pure white, bright red, rich crimson, and variegated colors. The flowers are mainly solitary on the armpits of the leaves, and in most varieties are very pretty. The plants are easily propagated from seed, cuttings, or divisions of the roots.

Above we present a beautiful engraving reproduced from a foreign journal, which shows a growing plant, together with a separate flower and seed capsule, both enlarged to show the form. This is taken from a new variety, *Antirrhinum Morea*, lately found in the Morea (Greece). Though not yet introduced in this country, it doubtless soon will be, and take a favorable rank with the new, improved seedlings of the present day. It is of elegant form, as the engraving shows, and is described as bearing flowers of lively yellow color which continue in succession during several weeks. Its fine cut leaves, and down-colored slender stems add greatly to its beauty. (1859)

Sweet Peas—Sow Early

There are some old-fashioned flowers of such real merit that they will always be popular. Among these is the sweet pea *(Lathyrus odoratus)*. Its beauty in the garden is sufficient to commend it, but as a cut flower it has especial merits, in its delicate colors, beauty of form and most exquisite fragrance; beside these it is remarkably lasting when cut.

As with the edible peas, our dry and hot summers make these of short duration with

us, but we can have them in much more satisfactory condition than is usual by observing two points: to sow them early and to sow them deep. Put in the seeds the very first thing after the soil is in proper condition, and let them be at least four inches below the surface. A good plan is to open a drill four inches deep, drop the seeds about two inches apart, and cover with an inch or so of soil; when the shoots begin to break ground, put on soil, a little at a time, until the drill is filled. The object of this is to place the roots well below the surface where they will not be so soon affected by the heat and drouth, and thus prolong their season of bloom. The vines must have supports of some kind; if sown near a fence, strings leading from a peg in the ground to a nail on the fence will answer. A very pleasing method is to sow the seed in circles, two feet in diameter; set in the center a stake about five feet high, and lead strings from pegs in the circle to the top of the stake. At one time, needing a low screen or hedge in the garden, we made it as follows:

Stout stakes were set at intervals and about five feet out of the ground; a few inches from the ground and at the tops of the stakes, string pieces about three inches wide were nailed on; pea brush was then set, not very closely, along and next to this trellis. The upper ends of the brush were brought close to the top rail by winding twine around the rail in such a manner as to include and hold the tops of the brush. The ends of the brush were, by use of the pruning shears, cut level with the top rail. The peas were sown at the base of this trellis and soon covered it, making a beautiful sight. There are now a number of fine varieties of the sweet pea, as may be seen by consulting the seedsmen's catalogues. (1881)

The Violet—Its Perfume

"The forward violet thus did I chide:
Sweet thief, whence didst thou steal
thy sweet that smells,
If not from my love's breath?"

The perfume exhaled by the *Viola adorata* is so universally admired that to speak in its favor would be more than superfluous. The demand for the essence of violets is far greater than the manufacturing perfumers are at present able to supply, and, as a consequence, it is difficult to procure the genuine article through the ordinary sources of trade.

Real violet is, however, sold by many of the retail perfumers of the West End of London, but at a price that prohibits its use except by the affluent or extravagant votaries of fashion. The true smelling principle or essential oil of violets has never yet been isolated; a very concentrated solution in alcohol impresses the olfactory nerve with the idea of the presence of hydrocyanic acid, which is, probably, a true impression. Burnett says that the plant *Viola tricolor* (heart's ease) when bruised, smells like peach kernels, and doubtless, therefore, contains prussic acid.

The flowers of the heart's ease are scentless, but the plant evidently contains a principle which, in other species of the viola is eliminated as the "sweet that smells," so beautifully alluded to by Shakespeare.

For commercial purposes, the odor of violet is procured in combination with spirit, oil, or suet, by maceration, or by *enfleurage*, the former method being principally adop-

ted, followed by, when "essence" is required, digesting the pomade in rectified alcohol.

A good imitation essence of violets is best prepared thus:

Spirituous extract of cassie pomade, 1 pint; esprit de rose from pomade, ½ pint; tincture of orris, ½ pint; spirituous extract of tuberose pomade, ½ pint; otto of almonds, 5 drops. (1856)

AQUILEGIA.

COCKSCOMB.

CYCLAMEN PERSICUM.

ECHEVERIA SECUNDA GLAUCA.

GOLDEN TRICOLOR GERANIUM.

"12 BEST" FUCHSIAS.

PANSY (*Type*).

Floral Vocabulary

ASTER, QUILLED.

MONTHLY CARNATION, "PETER HENDERSON."

Acacia, Yellow . . . . Concealed Love
Acacia, Rose . . . . Elegance
Acalea . . . . Temperance
Acanthus . . . . The Arts
Aconite-leafed Crowfoot . . . . Luster
Agnus Castus . . . . . Coldness without Love
Agrimony . . . . Thankfulness
Alyssum, Sweet . . . . Worth beyond Beauty
Althea Furtex . . . . Consumed by Love
Almond . . . . Hope
Aloe . . . . Bitterness
Ambrosia . . . . Returned Affection
Angelica . . . . 'Inspiration
Amaranth . . . . Immortality

Arum . . . . Ferocity and Deceit
Asphodel . . . . My Regrets follow you

Bachelor's Button . . . . Hope in Misery
Balsam . . . . Impatience
Balm . . . . Social Intercourse
Barberry . . . . Sourness
Basil . . . . Hatred
Beech . . . . Prosperity
Blue Bell . . . . Constancy
Bay Leaf . . . . I change but in dying
Bay Wreath . . . . The Reward of Merit
Betony . . . . Surprise
Bindweed . . . . Humility
Birch . . . . Gracefulness
Black Poplar . . . . Courage
Black Thorn . . . . Difficulty
Borage . . . . Bluntness
Box . . . . Stoicism
Broom . . . . Neatness
Buck Bean . . . . Calm Repose
Burdock . . . . Importunity
Buttercup . . . . Ingratitude

Calla . . . . Feminine Modesty
Calycanthus . . . . Benevolence
Candy Tuft . . . . Indifference

AMARANTHUS TRICOLOR.

Fig. 1.—CHESTNUT LEAVES AND FLOWERS.

Amaranth, Globe . . . . Unchangeable
Angrec . . . . Royalty
Anemone . . . . Frailty
Apple Blossom . . . . Fame speaks him
good and great
Apocynum . . . . Falsehood
Ash . . . . Grandeur
Aspen Tree . . . . Sensibility
Aster . . . . Beauty in Retirement
Amaryllis . . . . Beautiful, but timid
Auricula, Scarlet . . . . Pride

CANDYTUFT, DARK PURPLE.

Canterbury Blue Bell . . . . Gratitude
Cardinal's Flower . . . . Distinction
Carnation . . . . Disdain
Catchfly . . . . Artifice
Cedar Tree . . . . Strength
Chamomile . . . . Energy in Adversity
Cherry Blossom . . . . Spiritual Beauty
Chestnut . . . . Render me Justice

China Aster . . . . Variety
China Pink . . . . Aversion
Chrysanthemum . . . . Cheerfulness
Clematis . . . . Mental Beauty
Columbine . . . . ½ Folly
Coltsfoot . . . . Maternal Care
Coriander . . . . Concealed Worth
Coreopsis . . . . Ever Cheerful
Cowslip . . . . Native Grace

42

Crocus . . . . Youthful Gladness
Crown Imperial . . . . Majesty
Cypress . . . . Mourning
Daffodil . . . . Delusive Hope
Dahlia . . . . Dignity and Elegance

Daisy . . . . Innocence
Dandelion . . . . Oracle
Dew Plant . . . . Serenade
Dogwood . . . . Durability
Dragon Plant . . . . Snare

Eglantine . . . . Poetry
Elder . . . . Compassion
Elm . . . . Dignity
Enchanter's Nightshade . . . . . . Fascination
Evergreen . . . . Poverty
Everlasting . . . . Unceasing Remembrance

Fennel . . . . Strength
Fern . . . . Sincerity
Fir . . . . Time
Flax . . . . Acknowledged Kindness
Flowering Reed . . . . Confidence in Heaven
Flower of an Hour . . . . Delicate Beauty
Forget-me-not . . . . True Love
Foxglove . . . . I am ambitious for your sake
Fuchsia . . . . Confiding Love

Geranium, Nutmeg . . . . . . I shall meet you

IVY GERANIUM, "KING ALBERT."

Geranium, Lemon . . . . A tranquil Mind
Geranium, Oak . . . . True Friendship
Geranium, Rose . . . . Preference
Geranium, Scarlet . . . . Consolation
Geranium, Silver . . . . Recall
Geranium, Ivy . . . . Bridal Favor
Gilly Flower . . . . Lasting Beauty
Glory Flower . . . . Glorious Beauty
Golden Rod . . . . Encouragement
Grape, Wild . . . . Charity
Grass . . . . Utility

BROMUS BRIZÆFORMIS.

Harebell . . . . Grief
Hawthorn . . . . Hope
Hazel . . . . Reconciliation
Heath . . . . Solitude
Heart's Ease, or Pansy . . . . . Think of Me
Heliotrope . . . . Devotion
Hellebore . . . . Calumny
Holly . . . . Domestic Happiness
Hollyhock . . . . Fruitfulness
Hops . . . . Injustice
Horse Chestnut . . . . Luxuriancy
Horn Bean . . . . Ornament
House Leek . . . . Vivacity
Houstonia . . . . Content
Hyacinth . . . . Game, Play
Hydrangea . . . . Heartlessness
Ice Plant . . . . Your looks freeze me
Iceland Moss . . . . Health
Iris . . . . A message for you

IRIS.

Ivy . . . . Friendship
Jasmine, White . . . . Amiability
Jasmine, Yellow . . . . Elegant Gracefulness
Jonquil . . . . Desire

Judas Tree . . . . Unbelief
Juniper . . . . Protection
Kennedia . . . . Mental Excellence
King-Cup . . . . I wish I was rich
Larburnum . . . . Pensive Beauty
Lady's Slipper . . . . Capricious Beauty
Larkspur . . . . Fickleness
Larch . . . . Boldness
Laurel . . . . Glory
Laurustinus . . . . I die if neglected
Lavender . . . . Acknowledgment
Lemon Blossom . . . . Discretion
Lettuce . . . . Cold hearted
Lilac . . . . First emotions of love
Lily, White . . . . Purity and Modesty
Lily of the Valley . . . . Return of Happiness
Linden Tree . . . . Matrimony
Lobelia . . . . Malevolence
Locust . . . . Affection beyond the grave
London Pride . . . . Frivolity
Lotus . . . . Estranged Love
Love-in-a-mist . . . . Perplexity
Love-in-a-puzzle . . . . Embarrassment
Love-lies-a-bleeding . . . . Hopeless, not
heartless
Lucern . . . . Life
Lupine . . . . Sorrow, Dejection
Madwort, Rock . . . . Tranquility
Maize . . . . Plenty
Mallow . . . . Sweet Disposition
Magnolia . . . . Love of Nature
Mandrake . . . . Rarity
Maple . . . . Reserve
Marvel of Peru . . . . Timidity
Marigold . . . . Inquietude
Meadow Saffron . . . . My best days are past
Meadow Sweet . . . . Uselessness
Mercury . . . . Goodness
Mezereon . . . . Desire to please
Mignonette . . . . Excellence and loveliness

NEW HYBRID SPIRAL MIGNONETTE.

Mimosa . . . . Sensitiveness
Mint . . . . Virtue
Missletoe . . . . I surmount all obstacles
Moonwort . . . . Forgetfulness

43

Motherwort . . . . Secret Love
Moss, Tuft of . . . . Maternal Love
Mulberry Tree . . . . Wisdom
Mushroom . . . . Suspicion
Mouse Ear . . . . Forget-me-not
Myrtle . . . . Love in absence

WHITE CRAPE MYRTLE.

A CLUSTER OF NEW VARIETIES OF DRUMMOND'S PHLOX.

Narcissus . . . . Egotism
Nasturtium . . . . Patriotism
Nettle . . . . Slander
Nightshade . . . . Dark Thoughts
Night-Blooming Cereus . . . . Transient
Beauty
Nosegay . . . . Gallantry
Oak . . . . Hospitality

Fig. 2.—FLOWERS OF THE OAK.

Oats . . . . Music
Oleander . . . . Beware
Olive Branch . . . . Peace
Orange Tree . . . . Generosity
Orchis . . . . A Belle
Osier . . . . Frankness
Ox-Eye . . . . Obstacle

Palm . . . . Victory
Pansy, or Heart's Ease . . . . . . Think of me
Parsley . . . . Entertainment
Passion Flower . . . . . Religious superstition
Pea, Everlasting . . . . Wilt thou go with me
Pea, Sweet . . . . Departure
Peach-Blossom . . . . I am your Captive
Pennyroyal . . . . Flee away
Peony . . . . Ostentation
Periwinkle . . . . Sweet Remembrances
Peruvian Heliotrope . . . . Infatuation

Phlox . . . . We are United
Pimpernel . . . . Assignation
Pineapple . . . . You are perfect
Pine . . . . Pity
Pink . . . . Purity of Affection
Plane Tree . . . . Genius
Plum Tree . . . . Keep your Promises
Polyanthus . . . . Confidence
Pomegranate . . . . Foolishness
Poppy . . . . Consolation of Sleep
Prickly Pear . . . . Satire
Primrose . . . . Early Youth
Primrose, Evening . . . . I am more constant
than thou

PRIMULA SINENSIS FIMBRIATA.

Privet . . . . Prohibition
Pyrus Japonie . . . . Fairies' Fire
Petunia . . . . Thou are less proud
than they deem thee
Quamoclet . . . . Busybody
Queen's Rocket . . . . . . Queen of Coquettes
Ragged Robin . . . . Dandy

Rose Bud . . . . A Young Girl

Rose, Austrian . . . . Very Lovely
Rose, Bridal . . . . Happy Love
Rose, Bergundy . . . . Simplicity and Beauty
Rose, Damask . . . . Bashful Love
Rose, Monthly . . . . Beauty ever new
Rose, Moss . . . . . Pleasure without alloy
Rose, Multiflora . . . . Grace
Rose, White . . . . Silent Sadness
Rose, Musk . . . . . Capricious Beauty
Rose, Yellow . . . . Infidelity
Rosemary . . . . Remembrance
Rush . . . . Docility
Rue . . . . Purification
Saffron . . . . Excess is dangerous

THE SAFFLOWER (*Carthamus tinctorius*).

Star of Bethlehem . . . . The light of our path
Strawberry . . . . Perfect Excellence
Striped Pink . . . . Refusal
Sumach . . . . Splendor
Sun-Flower . . . . False riches
Sweet Brier . . . . Poetry
Sweet Flag . . . . Fitness
Sweet Sultan . . . . Felicity
Sweet-scented Tussilage . . . Justice shall be
done you
Sweet William . . . . A Smile
Syringa . . . . Memory

PERFECT TULIP.

Sage . . . . Domestic virtues
Scabious . . . . . . Unfortunate Attachment
Scarlet Ipomoea . . . . Attachment
Sensitive Plant . . . . Sensitiveness
Serpentine Cactus . . . . Horror
Snap Dragon . . . . Presumption
Snow-Ball . . . . Thoughts of Heaven
Snowdrop . . . . Consolation
Spider Wort . . . . Transient Happiness
Southern Wood . . . . Jesting
St. John's Wort . . . . Animosity

SWEET WILLIAM.

Verbena . . . . Sensibility
Vine . . . . Intoxication
Violet, White . . . . Candor
Violet, Blue . . . . Modesty
Violet, Yellow . . . . Rural happiness
Virgin's Bower . . . . Filial Love

Wall Flower . . . . Fidelity in misfortune
Wake Robin . . . . Ardor
Water Lily . . . . Purity of heart
Willow, Weeping . . . . Forsaken

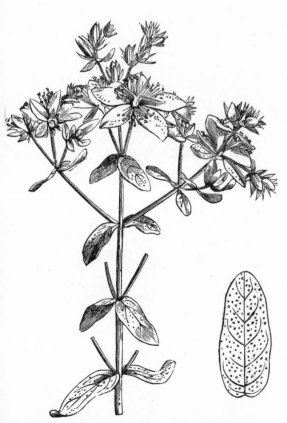

Fig. 1.—A FLOWER CLUSTER OF ST. JOHN'S WORT.

THE GREAT WHITE TRILLIUM OR "WAKE ROBIN" (*Trillium grandiflorum*).

Tamarisk . . . . Crime
Tansy . . . . Resistance
Teasel . . . . Misanthropy
Thistle . . . . I will never forget thee
Thorn Apple . . . . Deceitful Charms
Thyme . . . . Activity
Tremella . . . . Resistance
Trumpet Flower . . . . Separation
Tulip . . . . Declaration of love

Valerian . . . . Accomodating disposition
Venus's Looking-glass . . . . Flattery
Venus's Fly-trap . . . . Deceit

Wax Plant . . . . Susceptibility
Wheat . . . . Riches
Winter Cherry . . . . Deception
Witch Hazel . . . . A spell
Wood Sorrel . . . . Joy
Woodbine . . . . Fraternal love
Wormwood . . . . Absence

Yarrow . . . . Thou alone can'st cure
Yew . . . . Sorrow

Zinnia . . . . Absence
(1845)

III Cooking & Preserving

Brown Bread

Mrs. Henry Green, Saratoga Co., N.Y., sends the following which she thinks will be found superior to anything yet published in the *Agriculturist*. (We know that a very similar preparation *is* good.):

Mix 3 pints of sour milk or buttermilk, ½ cup molasses, 1 tablespoonful salt, 1 tablespoonful soda or saleratus, 5 cups of wheat or rye flour, and 5 cups of Indian meal. Put it in a pan, about 3 inches deep, and bake three hours in an oven heated as for wheat bread. (1862)

A First-Rate Corn Bread

It is hard to conceive, what, as a nation, we should do without our great staple Indian Corn, of which about a thousand million bushels, or more than fifty thousand pounds are now annually produced in our country. What crop would take its place? As an article of food it is both healthful and nutritious, and is hardly excelled even by wheat. Yet comparatively few families make any account of it in the culinary department.

The hasty-pudding or mush, poorly made, and not half boiled, the Johnny Cake (jour-ney cake?) made essentially of meal, salt, and water, or a little milk, are the chief articles of diet prepared from corn, in three families out of four. No wonder children grow up with a dislike of it, and in after life feel as an old gentleman remarked to us recently: "I don't want any 'Indian' in my family, I had enough of it while a boy."

But this should not be so; there is an almost infinite variety of wholesome, nourishing healthful preparations to be made of Indian corn, which are both pleasant to the taste, and economical withal. We have given several recipes for the preparations referred to, some cheap, and some more expensive, and we shall give many others. Here is one, partly meal and partly flour, which we have used for a few months, and which to our taste is first-rate:

To one quart of thick sour milk, or of buttermilk, add 1½ teacupfuls of molasses; 3 cups fine meal; 3½ cups of flours; 1 teaspoonful of salt, and 2 teaspoonfuls of soda. (Sweet milk may be used as well, by adding only 1½ teaspoonfuls of soda, and 3 of cream of tartar.) Stir well together, put into a basin, and steam three hours, then bake one hour. It is moist and delicious, and will keep good for several days. The steaming may be done in a regular, steaming vessel, or in any kettle, by simply setting the basin upon a brick block, to support it above the boiling water — just as bread is steamed. The kettle will need to be covered, of course. (1859)

Doughnuts Not "Greasy"

Here is an invention of our own which we might patent, but being employed to labor for the public, that public is entitled to our entire services. Everybody and his wife — and particularly his little folks — love the good old fashioned doughnuts, or "nut-cakes," or "crullers," or whatever name you call them. But many persons are troubled with weak digestion (dyspepsia), and the large amount of lard or grease absorbed by the said doughnuts does not always set well, but produces a rising in the stomach. When this is the case, try our invention.

The doughnuts being prepared as usual, just before immersing them in the hot fat, plump them into a well beaten egg. This will give a thin coating of albumen which will keep out the grease effectively. Furthermore, this coating will retain the moisture, and make them keep in good condition much longer than if not thus treated. (1859)

Plain Ginger Bread

To a cup of molasses add a piece of butter the size of a large walnut, the butter being melted, put in 1 cup sour milk, and a teaspoonful of soda. Spice with cloves or ginger; mix in enough flour to make a thick batter, and bake slowly. (1859)

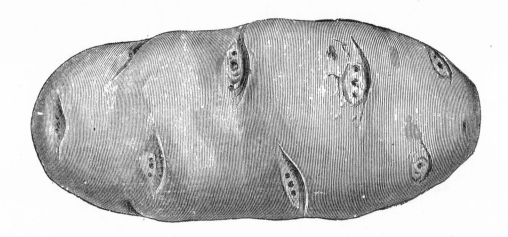

Lemon Cheese Cake

Contributed to the *Agriculturist* by Mrs. S. Wagstaff, Nebraska Territory. To 1 lb. of lump sugar, pounded, add 6 eggs, leaving out two of the whites, the juice of three lemons, the rinds of two grated, and ¼ lb. of butter. Put all the ingredients into a pan, and stir them gently over a slow fire until the mixture becomes thick, and looks like honey. Line the smallest size of patty pans with puff paste, put in a teaspoonful of the mixture, and bake. The mixture will keep twelve months, in a jar covered with paper, and set in a dry place. (1859)

Potato Bread

"Jeanne," of Erie Co., N.Y., writes: "We were glad to find in a former volume of the *Agriculturist*, directions for 'potato bread.' We have long been in the habit of using a few potatoes in bread, and think them an excellent addition, especially if the flour be dark or of inferior quality. For ordinary use we think our recipe better than the one given.

For five loaves of bread we select twelve nice white potatoes, and when cooking supper, boil them without breaking the skins if possible; then pour off the water, peel and mash very fine. Put with them a pint of cold water and stir in flour enough to make the whole a thick bottom. To this only lukewarm — avoid scalding it — add a teacupful of domestic yeast, or less if brewer's yeast be used. If kept warm over night it will be all in foam in the morning, ready to pour or sift through a colander. The sifting is facilitated by pouring in a quart or so of warm water while the colander is kept in motion with the other hand. Then stir and raise in a 'sponge' as in ordinary preparation of bread. It rises both in the sponge and in the loaf, much quicker than common yeast. This is a great improvement for biscuit, as it does not require half the 'shortening.' "

(*Remark* — Potatoes boiled and mashed so as to pass through a sieve or colander are without doubt a valuable addition to flour for both bread and pastry. The starchy, brittle character of the potatoes adds to the tenderness as well as the sweetness of the bread, and where potatoes are not as costly as flour, the addition is an economical one. The above mode may be adopted; or the sifted potatoes may be added directly to the flour, and the whole treated in the ordinary mode. Try it you who have not.—Editor) (1859)

Strawberry Shortcake

Having tried the article made after the following directions, furnished to the *Agriculturist* by Jennie V.V., of Queens Co., N.Y., we are ready to endorse them as first-rate — that is, for a shortcake.

To two teacupfuls of sour milk (water will answer where milk is scarce) add one teaspoonful saleratus; when this is dissolved put in one cup of butter or lard, and flour enough to make a soft dough. Roll it out into thin cakes, large enough to fill the pan in which they are to be baked. Dust a frying pan with flour, and bake the cakes over the fire, turning as soon as the under side is done, which will require but few minutes. Then split them open while hot and butter well. Have ready a quantity of strawberries well-sugared. Lay on a large dish a slice of shortcake, then a layer of strawberries, and so on alternately for five or six layers, and serve up — they will go down easily. (1859)

Strawberry Shortcake

If one partakes of strawberry shortcake in half a dozen different cafes, it is likely that a different preparation will be served at each. The old fashioned shortcake is in many cases replaced by a kind of confectionery made with slices of cake (somewhat like pound cake) covered with whipped cream, in which a few strawberries are imbedded. This is quite unlike the real thing. Among the recipes for shortcake that have been tried, the following

was preferred: flour, one quart; butter, three tablespoonfuls; buttermilk (or rich sour milk), one large cupful; one egg; white sugar (powdered), one teaspoonful; soda (dissolved in warm water), one teaspoonful; salt, one saltspoonful. Mix the salt and sugar with the flour; chop up the butter in the flour; add the egg and soda to the milk and mix, handling as little as possible. Roll out lightly, lay one sheet of paste upon the other in a round tin bake. While still warm, separate the cakes, and place between them a thick layer of strawberries, which should be abundantly sugared. Some place a layer of the fruit on the upper cake. It is eaten with sugar and cream. (1882)

To Make Plain Pie Crust

Take light bread dough sufficient to cover your pie plates and mix in butter, say a piece rather larger than a walnut to each pie. If sweet cream is at hand, 2 tablespoonsful added will be an improvement. Roll the crust out thin, and if you wish, spread on a little more butter and sprinkle with a little flour; then fold over and roll again; if rolled and folded several times it will be the better. Mashed potatoes mixed in the dough to make it seem short, are also an improvement. Indeed a most excellent crust can be made in this way, one which will not cause dyspepsia and one which dyspeptics can eat without injury. (1859)

A Novel Pumpkin Pie

Dear Editors: — I suppose you will not object to my talking with the good housewives who read the *American Agriculturist*. As I am a bachelor and must necessarily meddle with the household affairs, I deem it serviceable to study the mysteries of housekeeping. Some friends call me "Pot Betsy," and my fair city cousins think me over nice and fussy. Of course I don't care for what they say (?), and am very glad *this* house is comfortable and attractive enough to lure my most fastidious city friends to come and visit me.

Pray, madams, don't think that I do all the delicate work, for I employ my farmer's tidy wife, here. Now you know what I am, and will, I trust, allow me to tell you about a mysterious pie.

Yesterday I sat down *alone* with my venerable father at dinner, and noticed a nice pumpkin pie on the table, I was really surprised, because I was sure there was not a pumpkin in the house. We tasted it, and declared it to be made of pumpkins, but wondered where the housekeeper got them, and after dinner I asked her. She laughed and said it was made of turnips. — Turnips? Yes. I proposed to render her name famous by asking her recipe to be published in the *Agriculturist*, she smiled, pointed, and went out. Here is the recipe:

"Take two good sized yellow turnips; clean and peel them; boil about two hours and mash them. Then add 1 pound of brown sugar, 4 eggs, 2 quarts of milk, ½ cup of molasses, 1 spoonful of ginger, 1 nutmeg, ½ cup of wheat flour. This is then used the same as in making the real pumpkin, or custard pies. Mrs. L. E. Vail."

J.H.R., Passaic Co., New Jersey (1862)

Hints on the Art of Butter Making

In order to make pure butter, something is required besides the good breed of cows — the sweet grasses — the soft springs — the rolling lands — and the rich milk, and the most experienced churners — the most improved machinery — the purest atmosphere — the best material may be manufactured into yellow grease instead of butter, unless the process be properly performed. It is a fact well known to scientific dairymen, that the pure butter is not made by agitating the milk — not made by the process of churning. Butter already exists in the milk, and the art of separating it from the milk, is that on which the success of the dairy depends.

Butter exists in globules so small as to defy the detection of the eye unaided by the microscope, and the removal of these globules without crushing them is the delicate and difficult task the dairyman has

to do. There is no luxury that comes to the table which is so exquisitely sensitive as butter. If the cow feeds on white clover, the butter has a white clover flavor; if she feeds on cabbages, the butter has the flavor of cabbage; if the butter is kept in the vicinity of the stable, it forthwith becomes tainted with the smell of the stable; if packed away in pine tubs, it catches the taste and odor of the pine. It requires skillful handling or it will certainly be spoiled. If there is too much rubbing in the churn, these fine globules, mashed and crushed against the sides of the churn, will give greasy butter, and if the air is excluded the gases will injure it.

What can be done, you inquire, to cause the adhesion of the globules without grinding or breaking them. Experienced churners answer the question, when they caution young beginners not to churn too fast, not to heat the milk too much, not to overdo, etc. They may not in every instance understand the philosophy of the fact, but they do know the fact that "overdoing" makes grease and not butter.

The seasoning of butter is a matter of taste, and there are a great many persons who imagine that the more salt they put in the butter the better it keeps. That is a great mistake. Just enough and none too much is what is required. (Few indeed, even among butter makers in this country, know the luxury of *fresh* butter with no salt at all. — Ed.) Too much will spoil the taste and not save the butter.

Without penetrating any deeper at present into the philosophy of butter making, I will simply add that, a gentle and uniform agitation of the milk will best reward the butter maker for his pains. (The cream should have a temperature of about 65 degrees when churning begins. — Ed.) The butter should be kept away from all unpleasant odors, and when put down should be packed in white oak tubs.

Clean cows, clean stables, sweet churns and pans, neat and tidy operators, are among the things desired by those who would send pure butter to market. (1862)

Butter Molds and Stamps

H. M. Taylor, Kansas, asks whether there are any molds made by which butter may be put up in pound or half-pound cakes for the market. We give cuts of the usual forms of molds for this purpose. They are made of soft wood, as white ash or soft maple, and are generally kept for sale at all country stores where willow-ware is sold.

The manner of using them is as follows: When the butter is ready for making up, it is weighed out into the proper quantities, and each piece is worked in the butter dish with the ladle into flat round cakes. These cakes are either pressed with the mold shown in fig.

Fig. 1.—BUTTER-STAMP.

Fig. 2.—BUTTER-MOLD.

1, or are made to go into the cup of the mold shown at fig. 2. Inside of the cup, fig. 2, is a mold with a handle which works through a hole in the upper part of the cup. The cup is inverted on to the table, and when this handle is pressed down it forces the mold on to the butter, which is squeezed into a very neat ornamented cake. By pushing the handle and lifting the cup, the cake of butter is pushed out of the mold. This makes a very favorite mode of putting up fine butter for market, and is also well adapted for preparing butter for the table in houses where neatness of appearance is studied. The molds when in use should be kept wetted in cold water to prevent the butter from sticking. (1872)

Tomato Soup

A. D. Ferrer, Fergus, C.W., writes that a pot of soup even fit for Esq. Bunker, may be made as follows:

Take about two dozen ripe red tomatoes, a large teacupful of cream, with a good beef bone for a "strengthener," season with pepper and salt, and boil in sufficient water for two hours. (1859)

Cheap Pea Soup

Put into the iron pot two ounces of drippings, one quarter of a pound of bacon, cut into dice, two good onions sliced; fry them gently until brownish, then add one large or two

small turnips, the same of carrots, one leek, and one head of celery, all cut thin and slanting (if all these can not be obtained, use any of them, but about the same amount); fry for ten minutes more, and then add seven quarts of water; boil up, and add one pound and a half of split peas; simmer for two or three hours, until reduced to a pulp, which depends on the quality of the peas; then add two tablespoonfuls of salt, one of sugar, one of dried mint; mix half a pound of flour smooth in a pint of water, stir it well, pour in the soup, boil thirty minutes, and serve. (1862)

Cream of Celery

Celery is unusually scarce now, and little of it is well blanched and crisp. But even the somewhat inferior quality may be made quite palatable. Cut it into very small pieces, rejecting the toughest green portions. Add only water enough to keep it from burning, and boil it in a closely covered vessel for an hour, or until perfectly tender. Then add a sufficient quantity of milk, first thickened with a tablespoonful of flour to each pint, previously rubbed smooth with two tablespoonfuls of butter, and salt and pepper to the taste, very little of the pepper. Boil and serve as soon as the flour is thoroughly cooked. If made moderately thin with the milk, flour and butter, it can be rubbed through a colander, when it gives a delicious, cream-like soup. Smooth squares of bread well browned are frequently put into the soup when finished. A bowl of this, eaten with bread, the same as bread and milk, makes an excellent noon lunch. (1882)

Vegetable Soup

Take a good sized chicken, or an equivalent piece of beef or mutton, cut it up and put it in water, rather more than enough to cover it, adding a tablespoonful of salt; boil until nearly tender, skim off the fat; add butter, salt, and pepper, and more water if necessary; then slice into the soup ten large potatoes, one small Swedish turnip, one carrot, two parsnips, (an onion and a few stalks of celery,) with two or three spoonfuls of rice; boil half an hour, or until tender.

Before serving, add a spoonful or two of wheat flour stirred up with cold water. One or two spoonfuls of sweet cream greatly improves the flavor. (1862)

Blackberry Jam

Blackberries, in almost any form, and wine made from them, are very pleasant and wholesome, and besides are conceded to be good medicine for the "summer complaint," and housewives and nurses look out for a good stock of blackberry jam. It is easily made, and there is no difficulty in keeping it.

To each pound of ripe fruit, stewed in a porcelain kettle for five minutes, add 1 lb. light brown sugar, and mash the contents fine with a strong iron or wooden spoon while still upon the fire. When well mixed and boiled 15 minutes longer, stirring it well in the mean time, fill the scalded jars or glasses, and set away. (1862)

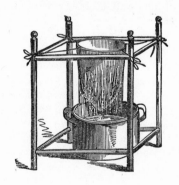

Jelly-stand,

Peach Jelly

Pare well-ripened peaches and remove the pits; boil the fruit until quite soft in water enough to cover it; strain through a coarse bag and add 1 lb. of white sugar to each quart of the liquid, boiling down until upon trial it stiffens when cooled. If it does not stiffen sufficiently, add a little isinglass. Put in jelly glasses, tumblers or bowls which have been boiled in water for several minutes. Cover with beeswax, and paste white paper over them. After setting a short time in the sun, preserve in a cool, dark place. (1862)

Salad Dressing

It is a great art to make a good dressing for green salad — lettuce or endive. The art consists in mingling the various ingredients so that each will become disguised and in its turn disguise the others while the combination in no wise obscures the delicious flavor of the lettuce or endives — but rather augments it and promotes its digestibility.

Take the yolks of two hard boiled eggs, crumble them with a silver fork or dessert spoon, add about half a teaspoonful of ground mustard and a teaspoonful of salt, and mix all well together. Then add in three portions a dessert spoonful of pure olive, walnut, or poppy oil, and rub the whole to a uniform smoothness. The addition of twice the quantity of oil will improve the salad to the taste of many, and nothing is more healthful; a dash of cayenne pepper, or a few drops of pepper vinegar may also be added; finally add about a dessert spoonful of sharp vinegar, and if the dressing is not fluid enough, a little water or more vinegar, adding it gradually and rubbing thoroughly all the time.

A little experience only is necessary, but it requires tact and patience. Most people abhor oil because they pour it over the green leaves instead of blending it in a dressing. Others douse on oil, catsup, mustard, pepper, salt — every spice or condiment they can lay hands upon — sugar it well besides, and then drown it in vinegar. Think of catsup on a crisp lettuce head — horrible! (1862)

Pleasant and Wholesome Summer Drink

The juice of currants, put up in airtight bottles, affords a foundation for a delicious and wholesome beverage. Put enough water with ripe currants to prevent their burning; heat in a preserve-kettle nearly to boiling; transfer to a bag suitable for straining and press out the juice; add half a pound of clean sugar to each pound of juice and boil, skim, and put up as recommended for putting up the fruit.

The juice alone will keep as well, if not better than the fruit, and, mixed with from one to two parts water, according to the taste, it makes a most refreshing drink. Being entirely free from alcoholic or intoxicating

properties, there is no danger of the creation of an appetite for strong drink resulting from its use. It might profitably be kept for sale by druggists at all seasons of the year, and we presume that putting it up for that purpose may be made a source of income worthy of the consideration of currant-raisers who now make wine. Those who usually have more of this abundant fruit than they know what to do with, may find it for their interest to take note of this suggestion. (1862)

Pot-au-feu Recipe Direct from France

Montargis, France
March 12th, 1862

Mr. Editor: Seeing in the January *Agriculturist* an inquiry for a recipe for the French "Pot-au-feu," I send one that is good, and at your disposal. Although so far away from our native land, we think we cannot do without your most excellent journal.

To 1 gallon of water, put 4 lbs. of beef, 3 teaspoonfuls of salt, and 1 of pepper; set it on the fire, and as the scum rises, skim it until clear. Then add 2 carrots, 2 turnips, 2 leeks, cut in pieces; 2 onions, in one of which stick 3 cloves; 1 burnt onion, or other coloring. Boil gently 5 or 6 hours. The broth, with good wheat bread, vermicelli, or tapioca, is good enough for any table. The meat is to be served afterward with the vegetables.

E.M.L.

Tomato Catsup—Tomato Sauce

The basis of tomato catsup, or ketchup, is the pulp of ripe tomatoes. Many defer making catsup until late in the season, when the cool nights cause the fruit to ripen slowly, and it may be it is gathered hurriedly for fear of a frost. The late fruit does not yield so rich a pulp as that gathered in its prime.

The fruit should have all green portions cut out, and be stewed gently until thoroughly cooked. The pulp is then to be separated from the skins, by rubbing through a wire sieve, so fine as to retain the seeds. The liquor thus obtained, is to be evaporated to a thick pulp, over a slow fire, and should be stirred to prevent scorching. The degree of evaporation will depend upon how thick it is desired to have the catsup. We prefer to make it so that it will just pour freely from the bottle. We observe no regular rule in flavoring. Use sufficient salt. Season with cloves, allspice, and mace, bruised and tied in a cloth, and boiled in the pulp; add a small quantity of powdered Cayenne. Some add the spices ground fine, directly to the pulp. A clove of garlic, bruised and tied in a cloth, to be boiled with spices, imparts delicious flavor. Some evaporate the pulp to a greater thickness than is needed, and then thin with vinegar or with wine.

An excellent and useful tomato sauce may be made by preparing the pulp, but adding no spices, and putting it in small bottles while hot, corking securely and sealing. If desired, the sauce may be salted before boiling, but this is not essential. To add to soups, stews, sauces and made dishes, a sauce thus prepared is an excellent substitute for the fresh fruit. It should be put in small bottles, containing as much as will be wanted at once, as it will not keep long after opening. (1882)

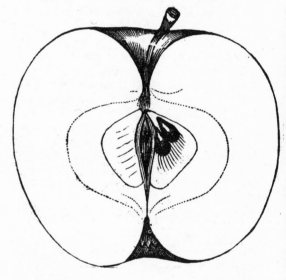

Good Apple Sauce

Contributed to the *Agriculturist* by Viola Homespun.

"Peel, quarter, and core as many apples as you wish to cook; put them in a tin or brass vessel with just water enough to cook them tender. While they are cooking, have a tin cup or some other small vessel on the fire, with about a half pint of water, one tablespoonful of butter, one of sugar, about ¼ of a nutmeg grated; when this boils, stir in enough paste (thickening) to make it of the consistency of cream; put your apples in a dish and pour this over them, and if you are fond of apple sauce you can't help liking this." (1862)

Vinegar Making

Vinegar, so useful in the household, is prepared from various materials, but whatever is used, or however the process of manufacture, its production in all cases depends upon the conversion of alcohol into acetic acid, or the acid of vinegar. Though the liquid used may not at first contain alcohol, it must have those principles from which it may be produced and alcohol is formed in the process before the liquid becomes vinegar. This is the case where fruit juices or solutions of sugar of any kind are used for vinegar; the change is first to produce alcohol from the sugar, and then to convert the alcohol so formed into acetic acid.

In apple, grape, and other fruit juices, no ferment is added as they contain a natural ferment, though vinegar is formed much sooner if some old vinegar, or mother of vinegar, be added. Vinegar prepared from fruit juices contains, besides acetic acid and water, various coloring matters, as well as peculiar flavoring principles; these, while they are not objectionable for table uses — indeed rather improve it — render it less fit for pickling, as the pickles have a less fine appearance and do not keep so well. Very pure and colorless vinegar is made directly from whiskey, or some other form of alcohol, and it is this which is found in the market as "white wine vinegar."

In making vinegar from alcohol a vat is used of the form shown in the accompanying figure. It may be either a vat built for the purpose or a very tall cask. They are made from 6 to 12 feet high, and we have seen the vats made of two casks put together, with the junction made tight by caulking. About a foot from the bottom of the vat are 6 or 8 half-inch holes, bored with a downward slant so that a liquid trickling down the sides of the cask will not run out, and an inch or two above the holes, a false bottom is placed in which are bored numerous ¾-inch holes. The cask is filled with beechwood shavings to within about a foot or 16 inches of the top. Six or eight inches below the top of the vat is fixed a platform, or cross partition in which holes are regularly placed, at 1½ inches apart. These are about 1-12th of an inch in diameter, and burned out so that they will remain free. This partition is put in place and the joint between it and the sides of the vat made tight by caulking. Pieces of twine are put into the holes of the partition in such a manner that the liquid, when poured upon it, will trickle through in drops. Four tubes of glass or of cane, ¾ of an inch in diameter, are set in holes in the partition; these do not project below, but above they reach to within

an inch of the top of the vat, which is closed by a tight cover having an opening to admit the liquid. A thermometer is inserted in a hole in the vat, 6 inches below the partition, so arranged that the internal temperature may be inspected. A wooden faucet is placed near the bottom of the vat, and a glass tube, curved in the form of a gooseneck, is placed with its bend below the row of air holes. The shavings are boiled in good vinegar before they are packed in the vat, and after all is ready, the vat is brought into fermentation by the use of a mixture of one-fifth vinegar and four-fifths of a 3 per cent mixture of alcohol and water.

This liquid is heated to 75 or 80 degrees, and poured into the vat and allowed to trickle through the shavings. The same liquid with the addition of more alcohol is warmed and passed through the next day, and so on until fermentation is well established, and the temperature within the vat has reached to about 100 degrees, when it is ready to commence the process of manufacturing vinegar.

The liquid used consists of 28½ gallons of water, 4 gallons of vinegar, and 10 quarts of 80 per cent alcohol. This, in passing through the vat, becomes converted into vinegar, and the process may be made continuous. In practice, two vats are used, and the liquid, with only a portion of the alcohol, is passed through the first vat, after which the remainder of the alcohol is added to it, and the process completed by passing it through the second. (1865)

Material for Pickles

There is a prejudice against pickles; perhaps it is because boarding-school girls of a sickly hue are said to dispose of marvelous quantities of them. Whatever the prejudice, it is not well founded. It is a blessed discovery that salt and vinegar will carry over something of the greenness of summer into the barren winter. Almost any vegetable preserved in good cider vinegar, is a healthful condiment, and aids digestion.

There is nothing better than the cucumber, and the vines are full of these in the summer. Pick them while small, and preserve in strong brine. Cabbage makes a good pickle, but we can keep this fresh through the winter, and use raw, which is better. Peppers, the thick-skinned squash variety, are almost indispensable in the pickle jar. We would suggest onions, were not the prejudice against and the love for this Egyptian esculent dispensable in the pickle jar. We would suggest onions, were not the prejudice against and the love for this Egyptian esculent universal. Then come mangoes, prepared from green fleshed melons, well stuffed with cabbage, horseradish, nasturtiums, white mustard and spices. This suits us a shade better than cucumbers. Beans in the pod, peaches not quite ripe, butternuts and hickory nuts, in the very green state — and last, but not least, tomatoes make good pickles. This list will enable the thrifty housewife to fill the pickle barrel. (1862)

Green Tomato Pickle (Sweet)

Contributed to the *Agriculturist* by E.E.J., Lisbon, Va.

"This pickle is very popular with us Virginians, and is thought to be particularly nice with mutton and beef, or any kind of fresh meat. Gather full-grown green tomatoes, scald and peel them. Make a strong ginger tea, into which drop your fruit and scald well. For every two pounds of tomatoes, take a pound of sugar and a pint of good vinegar, and make a syrup of this, and drop in the fruit. Let them cook until perfectly clear. Add cinnamon, mace, and white giner. Cover well with syrup, and tie up closely. (1862)

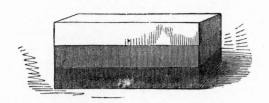

Ice Cream

This article is much talked about, and is supposed to be largely consumed in our cities; but the fact is, comparatively few persons know any\thing about genuine ice *cream*. Ice cream is chiefly made in cities and large villages — genuine cream "grows" in the country, and country people are the ones to have and enjoy the "simon pure article."

There have been two difficulties in the way: first, lack of ice; and second, the amount of apparatus and labor required. But ice houses are becoming quite common, so that in many places ice is always readily and cheaply accessible the year round. As for the apparatus, a good freezer is now got up so cheap, as to bring it within the reach of a majority of persons. The best freezer we know of is retailed as low as $3 each for the smaller (3 quarts) sizes. The freezer is complete in itself, requiring only the ice and salt to be put in — and the cream of course. This apparatus is so simple, and yet so philosophical, that a description will be interesting.

Fig. 1 shows its outward form; the

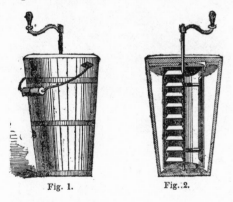

Fig. 1.   Fig.:2.

smallest size, for making 3 quarts of cream at a time, is somewhat taller than a large pail.

Fig. 2 shows the internal portions. The cylinder to hold the cream is so arranged that by turning the crank one way the cylinder itself is revolved in the surrounding ice and salt; while by turning the crank backward, only the wooden blades within are moved, including a scraper kept pressed against the cylinder by a spring, which removes the thin film of frozen cream formed on the tin. And just here lies the beauty of the invention, got up, not by a Yankee, by the way, but by a Pennsylvanian, and an editor at that, (H. B. Masser, editor of Sunbury *American*).

In the ordinary mode of freezing, the ice formed on the outside of the mass of cream,

acts as a non-conductor, and the internal portions are slow in congealing. In this freezer, the instant a thin film is frozen, it is scraped off and mixed with the whole mass. The wooden blades also keep the whole cream well beaten. The freezing is of course quickly performed, requiring little labor, and but little ice and salt around the outside of the cylinder.

The best, cheap, freezing mixture, is about one part of common salt to four parts of ice pounded very fine — if as fine as peas all the better.

For *cream*, good sweet cream, with sugar to the taste, and flavored with extract of lemon, pineapple, or vanilla, is all that is necessary. About 7 ounces of white sugar is required for a quart of cream. Those who can not get real cream may use, as a good substitute, sweet milk and eggs, well beaten together, say 2 eggs and 6 to 8 ounces of sugar to a quart of milk. Cook carefully for 20 to 30 minutes, then cool, flavor, and freeze.

THE ARCTIC CREAM FREEZER.

The manufacturers of the Arctic Freezer claim for it the following points, and are ready to prove them by public exhibition, if disputed.

1st. That they will actually freeze cream in four minutes.

2nd. They will freeze cream in less than one half the time, of any other freezer in use.

3d. They require much less ice than any other freezer.

4th. They will make cream smoother and lighter than any other freezer.

Sizes and Prizes: 3 quarts $3; 4 quarts $4; 6 qts. $5; 8 qts. $6; 14 qts. $8; 20 qts. $12.

A liberal discount to the trade.

E. S. & J. TORREY, Manufacturers, 72 Maiden Lane, New-York.

The following recipe is furnished by Mr. Masser, by which, he says, superior cream can be made for 18 cents per quart:

"Two quarts good rich milk; four fresh eggs, three quarters pound of white sugar; six teaspoons of Bermuda Arrow Root. Rub the arrow root smooth in a little boiled milk; beat the eggs and sugar together; bring the milk to the boiling point; then stir in the arrow root; remove it then from the fire and immediately add the eggs and sugar, stirring briskly, to keep the eggs from cooking, and then set aside to cool. If flavored with extracts, let it be done just before putting it in the freezer. If the vanilla bean is used, it must be boiled in the milk." (1859)

Snow Cream

Mrs. M. A. H. Rowe, Columbia Co., N.Y., says the following is quite equal to ice cream.

Beat thoroughly one egg with one cup white sugar, and one cup sweet cream, flavor to the taste, and stir in snow until it is quite stiff. (1862)

How to Make Ice Cream

This popular luxury may be enjoyed to perfection by our rural readers, who know what cream is, and who can readily obtain it. Here in the city, we get various substitutes for the genuine article, the best being made of milk and eggs, but much of that sold at the saloons is a compound of corn starch, arrowroot, and in some cases of ingredients known only to the makers. The readers will thank our kind contributor, Mr. J. Crozer, Mercer Co., N.J., for the following recipes and directions.

Vanilla Ice Cream. Take about one large vanilla bean to 3 quarts of pure cream. Split the bean; scrape the seeds into a cup; cut up the rest of the bean in fine pieces, and put with the seeds; add a little water, and let it stew for awhile over a fire; when done, and cool, add it to the cream. Add also, ½ lb. of fine white sugar for each quart of cream. When the sugar is dissolved, run the whole through a strainer into the can or freezer.

Lemon Cream. For 5 quarts of cream (equal to 9 qts. when frozen). Take 2½ pounds sugar, and about 3 lemons. Grate the outside of the lemons, and rub the gratings fine, with about 1 oz. sugar, then squeeze on to them the juice of two of the lemons, add a little more sugar, then mix with the cream. The cream should be sweetened before the lemon is added. Then strain into the freezer, and freeze.

Strawberry Cream. Use ½ pound sugar for each quart of cream, and strawberry juice for flavoring. The berries are squeezed through a piece of muslin, or a strainer, and an additional ½ pound of sugar is allowed for each pint of juice. Use only enough juice to give the cream a slight violet color. Confectioners add prepared cochineal, to heighten the color; still more juice can be added if desired. Place it in the freezer, etc., as for the others.

Pineapple Cream. Cut off the outside of a large ripe pineapple; cut up the rest in fine pieces in a pan, and cover it with sugar; or make alternate layers of the pineapple and sugar, and let it stand several hours; when wanted, squeeze it all through a strainer, and use sufficient to flavor—one pineapple to about 6 quarts cream. Allow, also, ½ pound sugar to each quart of the cream; and it is ready for freezing.

Almond Cream. Take ½ pound of sugar for each quart of cream; and about 1 ounce of bitter almonds to 6 quarts cream. Rub the almonds (which should be blanched) fine, in a mortar, or bowl, with rose-water. They should be prepared only as wanted, as they soon spoil. When fine, add them to the cream, etc. (1860)

Elderberry Wine

The following recipe, from an unknown source, has been tried by an acquaintance, and proved good:

1 quart juice, 3 quarts water, 4 lbs. sugar, and 1 tablespoonful yeast to the gallon. Put it in a cask, in a cool situation, keeping the cask full. It soon ferments and discharges froth from the bunghole. When the fermentation is over, bung up or bottle. Boiling juice injures it. (1862)

Rhubarb Wine

Take 4 lbs. of rhubarb to 1 gallon of water, squeeze it, put it into a tub, and pour the water on it; let it steep 3 days, then strain off the liquid; put 3½ lbs. of sugar to every gallon, and put it into a barrel, stir it every day for a fortnight, then add a few raisins and a small quantity of isinglass, then bung it up for three months. Finally bottle it, and in 5 or 6 weeks it will be ready for use. (1862)

Preserving Peaches

It may be done either with or without sugar, with much or with little. The question may well be considered, will sugar be cheaper next winter and spring than now, and we may act accordingly.

Peel and cut in quarters, put them directly into the bottles, with a very little water, put the bottles in a wash boiler, or similar vessel, filled with water to within two inches of the tops of the jars; bring the water to a boil, and boil it 15 minutes. Have prepared a syrup with 1 pound of sugar to a pint of water, or 1 pound to 2 quarts, just as you choose — the former usually preferable — pour this, boiling hot, into the bottles, as soon as they

are removed from the water, and close them immediately. (1862)

[After sealing, process in boiling water. Pints for 25 minutes; quarts for 30 minutes. —Ed.]

Keeping Apples

The ordinary method of stowing apples away in the bins of cellars is a very good one for family purposes, especially if the cellars be cool and dry in the warmer months, and of a temperature above the freezing point in winter. The best method, however, which we have found of keeping apples, is to pick them by hand from the trees in dry weather, as

soon as sufficiently ripe, and pack them in clean barrels, being very careful at the same time to prevent their getting bruised in so doing. Head them up tight from the air immediately, and place them in any cool place and dry, with the temperature as near 40 or 45 degrees as possible. In this way we have known them to remain perfectly sound for more than a year, and it is thus packed that they best bear transportation at sea.

As soon as we get a line of steamships to cross the Atlantic from New York to Liverpool in ten to twelve days, good apples will become quite an article of export; instead, therefore, of allowing their orchards to go to decay, as many, we are sorry to find, are doing, the production of good selected fruit should be more and more the study of the farmer, especially if he be the proprietor of only a small estate. Apples are undoubtedly worth raising even to be fed to pigs; and how much they contribute to the comforts and luxuries of the table, we need not say.

The varieties of apples to be grown on the farm need not be great; some 20 or at most 30 kinds for the summer, fall, and winter, would probably embrace all that are particularly desirable for family use. These should be well approved kinds, known as such by actual tests in our climate; for these greatly change by transplanting, not only from foreign countries, but even in our own diversified territories. We have repeatedly seen apples which were very superior in the northern and eastern states, prove quite ordinary on being transplanted to the west and south, and a knowledge of this fact should operate as a caution to those who purchase at

our nurseries, not to be over hasty in condemning everything which does not answer the description given it where first produced. (1843)

Buried Apples

"Experimentor," writing from Wyoming Co., Pa., says: "I have tried several plans for preserving apples but have never found anything that pleases me better than the old-fashioned Dutchess County way.

Take perfectly sound apples, and bury them in a dry place in the garden and keep them dry. Put 5 to 10 bushels in a heap; cover with about 3 or 4 inches of straw, and then with about the same depth of earth, leaving a small hole at the top, and then cover the whole with a straw cap. Or better with a roof of boards or slabs.

Apples treated in this manner retain their flavor, which is of the greatest importance. I have often, for late fall and early winter use, packed them in sawdust, wheat and oats chaff or wheat bran, dry sand, and in woolen blankets, etc., but none of these experiments have pleased me. They will not rot, but loose their fruity flavor." (1862)

Storing Potatoes

A uniform temperature of a few degrees above the freezing point, a moderately dry atmosphere, and exclusion from light, are the essential conditions for keeping potatoes safely through winter. How best to secure these depends upon various circumstances. Other conditions being equal, a cellar is always to be preferred for storage, yet immense quantities of potatoes are wintered every year in outdoor pits. When potatoes form one of the regular market crops of the farm, it pays to have a separate root cellar.

This need not necessarily be an expensive structure, but it must be so arranged that the floor is entirely dry at all times, that frost and

light can be excluded, and that complete ventilation can be provided when desired. Potatoes, to keep well, must be fully matured, should be dug when the soil is dry, and picked up soon after digging. Exposure to the sun and drying winds does not increase their keeping qualities. If at digging time the ground is sufficiently dry so that the tubers come out clean, they may be taken from the field directly to the cellar, provided this can be thoroughly ventilated. If the cellar cannot be kept cool during the fall months, it is better to store the potatoes at first in some dark out-building until the winter sets in earnest, when they are to be brought to the cellar.

The keeping quality of potatoes is seriously injured if they are kept too warm when first brought to the cellar. To prevent this the windows or ventilators should be kept open whenever the outside temperature is lower than that of the cellar, and closed when it is higher, the object being to keep the temperature as near the freezing point as possible without ever allowing it to fall below. A temperature of one or two degrees below thirty-two does not generally injure potatoes materially, especially when the cellar is dry, but if it falls below this point it certainly deteriorates their germinating if not their eating quality. It is not prudent, however, to run large risks in this direction. Therefore, upon the approach of severe cold weather the fore-handed farmer will make provision for emergencies which are sure to occur once in about ten years in the shape of extremely cold, penetrating winds, against which even so-called frost-proof cellars fail to offer sufficient protection.

Old carpets thrown over the heaps will protect potatoes against several degrees of frost. Straw or hay will accomplish the same object, but these are less convenient to handle and when brought in, the cellar doors have to be open so long as to increase still more the danger of freezing. An oil stove should constitute an indispensable adjunct to every root cellar. It costs but a few dollars, lasts a lifetime, and with an expenditure of a dime for oil its use during a very cold night may prevent the loss of a season's entire crop. Whenever the temperature in the cellar falls below the freezing point the stove should be placed in the coldest part of the cellar and lighted. It is astonishing how quickly the warming influence of even a small oil stove makes itself felt. In the absence of a stove a few lamps kept burning during cold nights are often sufficient to keep frost out of a small cellar, but half a dozen ordinary lamps do not give as much heat as a medium sized oil stove.

To maintain the proper degree of moisture in the atmosphere of a cellar is almost as important as the right temperature. If it is too dry the potatoes shrink and lose from ten to twenty per cent of their weight, and if it is too damp the tubers are apt to rot. The latter condition is the one most frequently met with, and to counteract it and drive out the

super-abundant moisture, the oil stove comes again into excellent service. By keeping it burning on a damp day for a few hours, while the ventilators or upper windows are open, the greater amount of moisture will be driven out of the cellar as if by magic. But few cellar floors are so dry that it is safe to place potatoes directly upon them, as the moisture rises from below and penetrates the entire heap. The easiest way to guard against this is to raise the bottom of the bins a few inches from the ground, so as to admit free circulation of air below them. (1888)

Storing Sweet Potatoes

In common with many farmers who save their own seed sweet potatoes, I have had more or less trouble in carrying the tubers through the winter in good condition. All of the usual methods have been enjoyed with varying success. The tubers have been wrapped in tissue paper, and stored in dark closets opening out of warm rooms; yet many rotted, or started so much that they would have been useless for bedding. I have laid them in rows about a chimney which was in daily use during the cold weather, and often have lost half of them.

A plan of handling and a place of storage, used last winter, gave me great satisfaction. The potatoes intended for wintering were selected for their good form and soundness, not one which had the slightest bruise being retained. They were laid as carefully as possible in baskets, from which they were taken one by one, and laid carefully away in a closet under a stairway, within three feet of the kitchen range. Having had more than were needed for seed in the spring, we used what we wanted for the table up to April, and the last ones were as good and well-flavored as they were in October. All were as plump as when put away, and not a single tuber went to waste. The closet was dark and the temperature was quite uniform at about sixty-five degrees. I ascribe my success in this instance, first, to the careful selection of sound specimens; second, to the care used in handling them, and, third, to the uniform temperature and dryness of the dark place of storage. (1893)

Keeping Cabbage in Winter

It is well known that freezing cabbages does not injure them materially, provided the frost be drawn out gradually, and it is a common

practice to bury them in pits or trenches out of doors, for keeping through winter. This answers a very good purpose, except that it is a rather troublesome operation to get at them when the ground is frozen hard, or covered with snow. Keeping them in the cellar of the dwelling is objectionable, as partial decay is induced by the warmth and dampness, and there is a most unsavory and unwholesome odor ascending to the rooms above; or if the cellar be very dry, the heads do not retain their freshness.

L. Bartlett in the *Boston Cultivator*, describes a method which he has used two winters past, by which these objections are obviated. He cuts off the stems, removes the loose outer leaves, and packs the heads in boxes or barrels with damp moss, such as is used by nurserymen in packing trees, roots, shrubbery, etc. These he has kept in the house cellar until March and in an outdoor cellar until late in Spring. Last winter, boxes so filled, were placed in the barn, and when frozen slightly were covered with straw, where they remained partially frozen, until April, without rotting or shriveling. When wanted for use, a head was placed in water an hour or two before boiling, and it then appeared as fresh as when taken from the field in November. He also suggests that clean straw cut fine and thoroughly wetted, might be successfully used instead of moss. It would be well worth while for market gardeners and others who wish to preserve cabbages, to make the experiment with straw upon a small scale; if successful it would be quite a saving of disagreeable labor. (1860)

IV   Home Care / Decoration

"FATHER IS COMING HOME!"

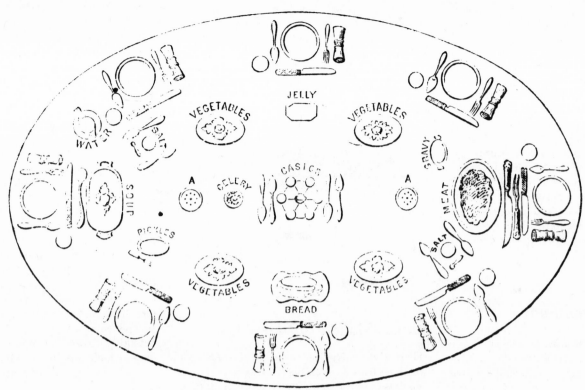

DIAGRAM FOR SETTING A TABLE.

Setting Out a Table

I have sketched a table on which I have arranged a simple dinner, in a style suitable for a family keeping one or two domestics, or none. I have placed the soup on the table with the meat, although if there be any one whose business it is to wait upon the table, it is better to have the soup served alone, the meat and vegetables being brought on when that is removed.

The lady of the house distributes the soup. It is not considered proper, as a general thing, to be helped a second time to soup. The soup plates should be placed before each person, and not in a pile by the tureen. As each one finishes his soup, his plate may be removed by the waiter, and a plate for the meat set before him. If there is no waiter, it is better for each one to retain his plate till all have laid down their spoons, and then one of the family can quietly put them aside.

There should always be regularity in the laying of a table. The dishes should not look like they had fallen down as hail stones, wherever it may happen. I have provided for four kinds of vegetables — if there are only two, they might be placed in the middle of the table.

Habits of eating are important, and no little straw shows more plainly which way the wind blows than these shown one's acquaintance, or want of acquaintance with society. When I was a child, I ate with my knife, and the great lesson was to teach me to put it to my mouth with the sharp edge from the lips. But now, in polite society, it is considered as great an offense against propriety to use the knife for any other purpose than to cut the food, as it then was to put it in the mouth in such a way as to be in danger of mingling my blood with my dinner.

When you have eaten all you wish, put your knife and fork side by side upon the plate, in close and loving union — with the handles at the right; and do not push the plate from you, but let it stand where you have used it.

Never use your own knife or fork to help yourself to salt, butter, vegetables or anything else. It is an abomination.

When you rise from the table, do not put the chair back against the wall, or push it under the table, but leave it where it is.

When jelly or sauce is used at dinner, it does not require a small plate, but should be put on the dinner plate.

Have the salts full, and the top nicely smoothed by passing a knife over it. Leave no salt scattered on the top of the glass. Be careful not to forget salt spoons.

Do not touch your hair while at table, nor pick your teeth — *and above all do not suck them.* That is enough to drive a person of refinement away from the table. It is worse than going round Point Judith, to hear such a sound. The very thought of it is nauseating. (1868)

Selecting Furniture, Arranging Rooms

Good taste may show itself quite as readily in a log cottage as in a Fifth Avenue palace, and be equally attractive. I remember no dwellings with more pleasure than some of the vine-covered log houses of the West, and a very simple little cottage in New Jersey, almost hidden in the loving embrace of roses, honeysuckles and grape vines. It does not require wealth to create home beauty. Refinement and delicacy of taste can invest the rudest home with charms that money alone can never furnish.

I have often been impressed with the wonderful sameness with which houses are furnished, as if, instead of consulting our individual wants and taste, we consulted the opinions and taste of others. A sofa of hair-cloth, and six mahogany chairs in black, seems to be considered almost essential where more showy and expensive furniture is not used, and then there must be a table with a marble top and a side-table or two, and a rocking chair, and these are too often arranged with mathematical precision against the walls, so as to destroy all idea of comfort and ease. Hair-cloth furniture, well made, is serviceable, but is very somber. A room furnished with it, unless relieved by bright colors in carpets and curtains, has a funeral look that is anything but cheerful. I much prefer cane or rush-seated chairs of prettier style. They are lighter to move, and more comfortable for use, than most stuffed chairs.

Furniture should be so arranged as to indicate that it is designed for use, and not merely kept for show. I should have a social, friendly air, as if on good terms with its neighbors and not afraid to meet on terms of

61

equality. Do not arrange books on a table as if they were paraded for military display, and ready to be marched around with measured step at just the same distance from the edge. Books, it is to be supposed, are to be read, and they should lie about carelessly, as if they had just been put down by a reader.

Do not select one article very much handsomer than the others. A velvet carpet calls for corresponding expense in sofas, chairs, and tables, while a pretty ingrain of good colors looks well enough for any country or city house of moderate pretensions. It is well to furnish a house so much within one's means as not to be constantly afraid that this and that will wear out, and can never be replaced. It is far better to be able to use and enjoy what we have and permit our children to enjoy it, too. (1857)

A Well-Arranged Pantry

In this age of multitudinous articles in the shape of china, glass and other ware used in cooking and at the table, a special room in the form of a pantry is needed for their proper care and protection. In the accompanying engraving is shown a portion of a pantry that is well arranged and convenient.

The window is placed at the end of the room, while around three sides is put a shelf, 30 inches in height, and from 15 to 20 inches wide. Beneath this shelf are placed drawers, cupboards, shelves, etc. The shelf situated in the end of the pantry is left unobstructed for cutting meat, bread, mixing pastry, and even for washing dishes. It will be found the most convenient and most used part of the whole room, being near the light, and with plenty of space. At either side of the pantry are arranged other cupboards, shelves, drawers, etc., in any manner thought most convenient, and are used for holding table linen, glassware, china, silver, etc., while those near the floor are for kettles, pans, pails, and other coarse kitchen ware.

Some may think ten or a dozen drawers are far too many for one pantry, but in practice,

A COMMODIOUS AND WELL-ARRANGED PANTRY.

more, instead of a less number, could be economically used. It is best to furnish some of the drawers with locks and keys, that they may be securely fastened when desired. Paint the woodwork in some light color; it is better to simply oil the top of the shelf near the window with raw linseed oil, as it is much used, and paint soon wears off. (1882)

Home-Made Fruit Dryer

There are inquiries before us for the cheapest and best fruit dryer. There are several patented dryers in the market, some of which turn out a superior quality of dried fruit, but they are somewhat expensive, and not within the reach of those, who have only a small amount of fruit that must be dried, without an outlay of money, or not at all. The accompanying engraving shows a form of dryer that is designed for indoor use, with the heat of a common stove.

The size of the frame is determined by that of the stove, and may be made of 1½-inch material. The legs should be long enough to support the frame well above the stove, and can be fastened on with hinges, so as to fold up when not in use. The frame may be either covered with a network of coarse twine or fine wire—wire cloth is best but more expensive. A second frame may be suspended below the first one, thus doubling the amount of the drying surface. (1882)

Convenient Quilting Frame

The quilting frames in ordinary use are an almost unmitigated nuisance. Except in the largest apartments they monopolize the room, and resting loosely upon the backs of chairs are frequently thrown down by a thoughtless urchin, to the great annoyance of the good housewife. The following plan, contributed by S. A. Newton, Susquehanna Co., Pa., remedies these inconveniences, and also gives a much better means for stretching and holding the quilt to its full tension.

The two bars A are from 7 to 8 feet long, or a foot longer than any quilt. They should be 2¼ inches thick, made eight square, and perfectly straight. A strip of cloth, b, is tacked to each bar, to which the quilt is to be attached. One end of each bar is fitted with a ratchet wheel, c. These ratchet wheels are attached to iron caps which fit upon the head of the bar. The pinion of the wheel has one end sharpened to insert in the bar, and the other extends outward through an opening in the horse which supports the bar. The other end of each bar has the cap and pinion without

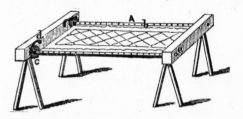

the ratchet wheel. The caps serve as bands to prevent the ends of the bars from splitting.

The horses are made of convenient height, with a sufficient spread of legs to stand firmly. Two dogs, d, are attached to one of the horses, to work in the ratchet wheels, and hold the quilt in place when stretched. The bars of the horses should be just long enough for quilters on opposite sides to reach over the quilt bars and meet half way. As fast as a section of the quilt is finished, another part is unrolled by lifting the dogs and rolling the bars, until the whole is completed. If ratchet wheels made of iron for the above arrangement can not be easily procured, and fitted directly upon the ends of the bars, which could be reduced in size to work in inch holes in the horses. (1862)

Quilts & Quilting Frames

It looks like a waste of time to buy calico, cut it up into inch pieces, and spend the leisure evenings of a whole winter in sewing it together again in fantastic shapes, to make a bed quilt, which, when finished, often looks like a failure to harmonize the chaos of a rag bag. There may be economy in using remnants of cloth and partly worn garments in this way, and much taste may be displayed in the arrangement, but quilting merely to make a spread in the manner first described, is not profitable industry. Making quilts of pieces otherwise wasted is a good initiation into the mysteries of needlework for the younger girls, and quilts not too thickly wadded, make very good bed clothing. Quilting parties, too, are very popular social gatherings, promotive of pleasant feeling among neighbors, and too fully Americanized to be easily abolished. Quilting frames, however, are generally voted a nuisance by the men folks. As usually made, with four ten-feet strips, and set upon the backs of chairs, they monopolize the room occupied for the time, are easily overturned, and occasion much inconvenience.

As many have by this time finished some patchwork, we will try and aid them in completing the quilting by giving a plan of a very simple frame, occupying much less room than the ordinary kind. It was furnished by Isaac Evans, Crawford Co., Ill. The men folks will no doubt be glad to make a set, both to please their better halves, and to abate the nuisance they complain of. The end pieces are made only about four feet long, with an open mortise let into each end to receive the rounded extremity of the side pieces. Holes are bored through at right angles with the mortises, to receive pins, with which to keep the stretchers or long side pieces in place, when the frame is in use. The long pieces may be of plank strips, left square, or better, worked eight-sided the whole length, with the ends rounded to turn easily in the mortises. Holes are let through to receive the pins that pass through the mortises. When wanted for use, the quilt is stitched to cloth strips fastened on the side pieces in the usual manner. One side is rolled up, the ends of the stretchers are placed in the mortises, and fastened with pins, as shown in the cut, and all is ready. It is better to have legs attached to the end pieces, on which the frame may stand, than to rest it on the backs of chairs. When one width is finished, and the quilt is ready to roll, the pins are removed, the quilted part brought on to the front roller by turning the side piece, and the pins inserted again. (1880)

Hints About Making Rag Carpets

Mrs. May Stuart Smith, of the University of Virginia, sends us the following:

Making rag carpets is one of the things that should not be, and doubtless will not be relegated to a place among the lost arts. Such carpets supply to many families comfort that would be unattainable otherwise. If the housewife is too busy to make them herself, there are always poor unhandy women without capacity for higher service, who are very glad to rip, cut and tack pieces together for the smallest remuneration, and at ten cents a pound, they can earn some part of their scanty livelihood and are glad to do it, and we are pleased to give them the work.

Such carpets turn to useful account many a fragment which would otherwise be thrown away, old furniture covers, garments, etc., without strength enough for any other good purpose. Cotton, woolen or silk, come into play equally well, if they are barely strong enough to be cut and wound into balls. Half an inch wide is the general rule, a little wider if rotten or very thin, and narrower for thick woolen stuff. The narrower they are cut the further they will go in making length, of course at the expense of thickness.

A pound of rags to the yard is the general calculation, though one of the nicest carpets I have seen, was a yard and three-quarters to the pound. For warp, bale cotton No. 6, containing thirty hanks to the bunch is best, though No. 7 will do and goes further. The weaver's allowance is two hanks to the yard for nearly yard wide carpet, but two hanks should be allowed for cross threads. Two bunches of No. 6 will therefore make 28 yards of woven carpet, and of No. 7, 44 yards. Warp doubled and not twisted, will be found to wear best.

Here is an important point for the inexperienced. In weaving, have a shuttle filled with the warp cotton and weave in two threads between the strip of rags. It will greatly strengthen the carpet, and if the warp is of fresh strong cotton, the carpet will wear until you are perhaps tired of it, or rich enough to buy a more costly one. I doubt, however, whether any other carpet ever gives you so much real pleasure as the rag one which was the result of your own contrivance and patience.

To color the warp, divide it into two equal portions, and with a package of dye, dye half black, using as a mordant a little copperas and bluestone. Dye the other half with copperas, afterwards washed out with weak lye. Have it put on the loom in stripes not over four inches wide; check it with the filling as tastefully and systematically as possible to form exact squares with the warp stripes. All the old flannel and white pieces may be put together and dyed to make the carpet brighter.

A simple way of dyeing beautiful red colors is to get red aniline from the drug store, tie it in a thin muslin bag, soak it in cold water to be afterwards added to a large kettle of hot boiling water in which a tablespoonful of alum is dissolved. Wet the pieces well and dye as much as your kettle will conveniently hold. This dye will keep and may be used more than once. Green and blue aniline are

first dissolved in a little alcohol, and then used the same as the red. In the country, walnut bark and the nut shells, red oak, pine and walnut bark, or sumac berries, are used, in each case dipping the fabrics in lye afterwards to set the coloring matter.

Another style of carpet is said to look very well, and is certainly easier to manage, viz.: have no design or stripe or check, but with the warp dyed any good serviceable color, weave in the pieces of all colors sewed together in such a way as to diversity them as much as possible. (1884)

A HOME-MADE HAMMOCK.

Make Your Own Hammock

A pretty and very comfortable hammock can be made of awning cloth. Two pieces, six feet long and a little over a yard wide, are cut for the body of the hammock; and two strips, five inches wide, to go along the sides. These are scalloped and bound with worsted braid, and the strips basted in place between the two large pieces. The side seams are sewed up on the wrong side. After being turned right side out, the two ends are bound with braid. Eight curtain rings are put on each end, and to each ring is fastened a heavy hammock cord. These cords are all joined to a large iron ring.

To hang the hammock, a light rope is passed through the rings and around two trees or posts. It will be more comfortable if the head is hung some inches higher than the foot. (1883)

Summer Awnings

It is very desirable to have some kind of protection over doors or windows which are exposed to the glare of the sun in summer. In a new place the trees are seldom large enough to afford much shade, while a porch is often a thing of the future. An elaborate awning with patent iron frame, etc., to be lowered and raised at pleasure, differs very materially in cost from the one here represented. The latter can easily be made at home.

It has only a stationary wooden frame, it is true, but it answers the purpose just as well as a more expensive and pretentious affair. A square frame is made of strips of wood, and fastened firmly to the side of the house, a little above the door or window, with another

AN AWNING FOR A DOORWAY.

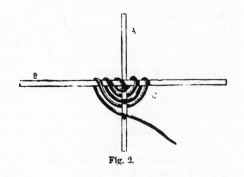

Fig. 2.

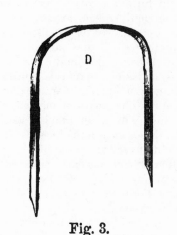

Fig. 3.

wooden strip reaching from the house to each of the front corners of the frame, to hold it in place. The awning is wide-stripe bed ticking (or regular awning cloth, if preferred), the top and front all in one. This requires two or more widths of material, and the side pieces are cut to fit, and sewed on. The lower edge is scolloped out all round, and bound with bright colored braid, and the whole is tacked closely and firmly to the frame. This awning is durable as well as ornamental, and need not be taken down all summer. (1883)

Home-Made Baskets

I observe your call for information as to how willow baskets are made. Having often, when a boy, seen my father's plowmen make baskets for farm purposes during the long winter evenings, I will endeavor to tell you how it was done.

The willows were never peeled, but were soaked in water in a long pig trough. The hoops, ribs and handles were generally made of split ash. Two ash splints were bent into hoops and placed one within the other, but at right angles, as shown in figs. 1, a and b. If

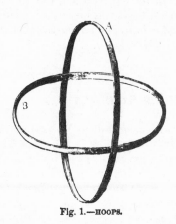

Fig. 1.—HOOPS.

the two hoops are round, of the same size and cross in the middle, the basket will be round — that is hemispherical. If the inner hoop, which forms the top of the basket, be the larger, the basket will be more or less oval. Having arranged the hoops, take the willow rods and weave them, as in fig. 2. As the semicircle enlarges, insert the ribs, bent as shown

in fig. 3, and of such size and form as to regulate the shape or contour of the basket, as shown at d,d, fig. 4. The splints, if well soaked, can be easily bent into any shape. Weave willows among these ribs or hoops — it being simply done — out and in, out and in, until the semi-circle reaches down the side of the basket to about the dotted line, fig. 4. Then commence on the ends of the basket and weave willows down the ends and along the bottom, as seen on the right hand end at e, fig. 4, putting in more of the ribs, fig. 3, if the work is large and consequently open.

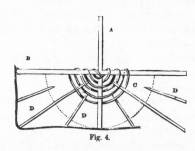

Fig. 4.

When you have the ends and sides well woven, the basket is finished, so far as strength is concerned. You can not pull it apart. But there will be gores or vacant spots left. Fill these up with weaving, and your basket is done. It matters little where a willow ends, provided it does not end at the top without going round the upper hoop. The wet willow twigs will bend easily and take any form, and yet when they dry they will be hard and stiff, and not be unwoven. Push several layers of willow close together and pull them tight; they will hold very firmly.

After the basket is dry, trim off all straggling ends with a sharp knife. A little practice, and perhaps the spoiling of a few baskets made with materials that are quite cheap, will enable you to make an article worth a dozen of what you can buy. Do not give it up if you cannot make a neat job at first. A little care and patience, with a modicum of practice, will set you all right. (1867)

Home-Made Corn Brooms

Broom-corn is grown just like common corn, and requires similar soil and cultivation. The seed should be planted thicker to avoid failures, and many prefer to put it in drills. When the seed begins to harden, the upper joint is broken over to prevent the head from breaking off the stalk too short. Before hard frosts, the heads with about a foot of stalk are cut off and laid on the barn floor, or on rails or poles under cover to dry thoroughly.

To clear off the seed and prepare the brush, a wooden comb may be made with a saw, on the end of a board, like fig. 1. The board may be nailed against a log or bench, or be cut short and secured in a workbench vise, if one be at hand. Save the seed for horses or poultry, selecting first, enough good plump kernels for the next planting. Cut off the stems about six inches from the brush.

Fig. 1.

When ready to use, take as much as is needed, and set the stalk portions in water up to the brush, and leave to soak an hour or two. When softened, gather in the hands enough for a broom, with the largest and best stalks on the outside, in regular order. The good appearance of the perfected broom will depend upon the evenness of the brush and good arrangement of the outside layers. Next, fasten a strong small cord to the ceiling, with a loop for the foot in the lower end, or tie a stick in, upon which to place the foot. Wind this cord two or three times around the brush as shown in fig. 2. Grasp the brush firmly in both hands, and roll it round several times, increasing the pressure with the foot. The next operation is to wind on a strong twine for a space of 1½ or 2 inches. This is best done by rolling the pressing cord to the part next to the brush, wind the twine on, and roll off the cord towards the end, following it with the twine. To make a neat knot at the end, double one end of the twine and lay it along the outside of the stalks as shown in fig. 3, letting the loose end lie out

Fig. 2.

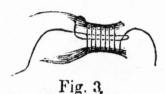

Fig. 3.

at the left. When the twine is all on, slip the right end through the loop, and draw the left end so as to bring the loop in under the coil of twine; then cut off the two ends close in to the coil. No knot will now be visible, as the lock is out of sight, and the ends are securely fastened.

If a flat broom is to be made, which is desirable, press the brush part between two narrow boards, fig. 4, fastened near together at one end with a piece of leather nailed on. The other end of the boards may be held together with a string. Instead of these boards the brush may be put between two short boards and screwed into a vise. The sewing is next in order. For this, use a large needle of iron or steel, or one of strong wood will do, fig. 5, rigid to twelve inches in length. At the

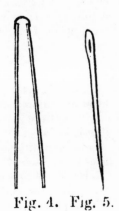

Fig. 4.   Fig. 5.

point where you wish to fasten the brush portion, say three or four inches below the winding thread, wind a twine once, or better twice around, and tie it firmly, leaving enough of one end to sew with. Now sew

through and through, letting the thread pass around the winding as shown in fig. 6. Point the needle forward in making each stitch, so as to have it come out on the opposite side a little further along each time. A second sewing may then be made further towards the lower end. Three sewings are sometimes made. Two will generally be enough, except where the brush is very long.

Sharpen the lower end of the handle, and drive it in exactly in the center, and fasten it with two small nails upon opposite sides, and the broom is complete. The lower ends of the brush may need clipping a little to make them even. With a little practice a very neat broom may thus be made. They may be made still more tasteful, though not stronger nor more durable, by using wire instead of twine, and by paring down the stalks, so as to make a smaller, neater shank. (1860)

Fig. 6.

How Candles Are Made

It seems probable that we shall have to go through a period when every economy must be practiced. Many of the little economies which were formerly common in farm houses, will come in fashion again, and the use of homemade candles in place of "store" candles will return. Some young housekeepers never knew how candles were made, and would like to learn, while some of the older ones, who have forgotten, wish to be reminded.

If clear tallow is used, the candles will crack and break. This is because it is too hard, a quality depending much upon the manner in which the animal has been fed. Dry food, such as hay and corn, produces very hard fat, while with grass, turnips and linseed meal, the fat is softer, and will not crack when made into candles. For this reason summer-made tallow needs no addition to soften it, but winter tallow should have one-eighth part of lard added to it to make it more suitable for candles. The easiest manner of making candles is to melt some tallow in a shallow, broad pan, such as a 10-

Fig. 1.—CANDLE WICKS.

quart milk pan, and having some wicks prepared to string them upon a wire or a round stick, see fig. 1, and holding them over the pan of tallow which is kept warm, but not hot, upon the stove, to dip up the fat with a spoon, and pour it down the wicks a few times until some of the tallow has hardened upon them. Several of these sticks full of wicks are prepared, and are taken one by one, and treated to a coating of tallow. As each one is greased, it is placed upon a frame, made as shown in fig. 2. The wicks are cut to the proper length, which is about 18 inches, and doubled. They may be measured to the proper length on a very simple frame, made by setting a rod upright at the end of a piece of board, a foot long, and fixing a bent wire at a proper distance from the rod. (see fig. 3) The wick is doubled around the upright rod, and as it is held on to the wire, it is cut with scissors by some one who helps. The wicks are left upon the rod until a quantity are ready, when they are put upon the sticks, as in fig. 1.

Fig. 2.—HANGING THE CANDLES.

Fig. 3.—WICK MEASURE.

Another and better plan is to make a narrow wooden box with a lid, and a sheet-iron or a tin bottom, turned up an inch or two, and tacked close (fig. 4). This is set

Fig. 4.—BOX FOR MELTING TALLOW.

upon the stove, and the tallow melted into it. It is deep enough to permit the wicks to be put into them the whole length. The wicks strung upon the wicks are dipped into the tallow three or four times slowly, until no more fat will harden upon them, when the bottoms of the wicks are touched upon the sloping lid, to take off the drip, and the stick full is hung upon the frame to cool, (fig. 2). Another is immediately taken, and treated in the same manner, and so on, until the first one comes around again. When there is enough tallow upon the wicks, the candles are finished, and appear as at fig. 4. These are the common dip candles.

The molded candles are made by pouring the tallow into a frame of molds, made of tin, as at fig. 6. The wicks are cut the proper

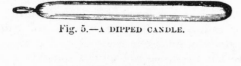

Fig. 5.—A DIPPED CANDLE.

Fig. 6.—CANDLE MOLDS.

length, strung upon thin sticks, and are inserted into the molds, the ends being drawn tight, and a knot is made to hold them in their place. They are set exactly in the center of the molds. Then the tallow is poured into the molds, until they are full, and they are hung up in a cool place for the candles to harden. The tallow shrinks as it cools, and the molds are made slightly tapering, so that it is easy to draw the candles out by the stick, on

which they are strung. The candles may be tied together by the loops at the ends in bunches of six or eight, and hung up in a cool, dry place. Candles thus made need to be snuffed when burning, which is a troublesome thing to do. If the wicks are plaited and flattened out before being used, they will bend over to one side when burned, and as the charred portion becomes exposed to the air outside of the flame, it wastes away, falls, or is carried off by the upward draft. The guttering of candles is caused by the tallow being too soft, or the wicks being twisted too hard; in either case the melted tallow is not absorbed by the wicks as fast as it becomes fluid, and so flows over the edge. (1876)

Experience in Soap Making

Whatever may be said about the advantages of selling ashes and grease and buying soap, it is best for most living in a farming community to make their own soap, and in a new country there is no alternative.

I start the lye to boiling, and then while boiling, if the lye is not strong enough to eat the feather off a quill, boil it down until it is. When it will just eat the feather, let the kettle be a little more than one-third full of lye, and put in grease, skins of the hogs, bacon rinds, meat fryings, and the like, until the kettle is about two-thirds full.

The kettle must not be full for with the least bit too much fire, over the soap goes. It is better to put in a little less than the necessary amount of grease. Lye and grease combine in certain proportions, but pass the limit, and no amount of boiling will take up an excess of grease. It will remain on top, hot or cold, and will be very troublesome; whereas a little too much lye will sink to the bottom when the soap comes.

If the proportions are good, a little fire only is required to keep it boiling, and in a few hours it is done. Then take a bucket of weak lye, and let it boil up with the soap once. This will not disturb the already made soap, but will wash the dirt out that was in the grease, and with it settle to the bottom. When the soap is cold it can be cut out in cakes. Exposure to the air will soften it down until it is about the consistence of mush, and little darker, growing fairer and fairer. (1869)

Straw Mats—How to Make Them

Straw mats are useful in every garden. They serve to cover half hardy shrubs during the winter, are handy to throw over tender plants to shield them from frosts, and where there

are hot beds, they are almost indispensable, to protect them during the cold nights of spring. They may be of any size, but it will be found most convenient to have them of a size to cover a single sash.

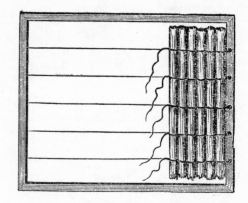

Make a rough frame one foot larger each way than the required mat; lengthwise of this, stretch pieces of large, strong twine, to serve as the warp, which may be tied to the frame itself or fastened to pegs placed in it for the purpose. The pieces of twine should be 8 or 10 inches apart, and the distance between the two outer ones about 6 inches less than the width of the mat. A piece of smaller twine 3 or 4 feet long is to be tied firmly to each thread of warp, close to one end of the frame; these are to serve as laces or binders to hold the straw in place. The frame being laid flat at a convenient height upon horses or in some other way, the workman stands inside of it facing the end where the binders are attached, he takes a small handful of straw and lays it with the butt ends projecting about 3 inches beyond one of the outer pieces of warp, and secures it by passing the second binder over it and tying this by means of a half hitch to the warp.

Another handful is similarly placed on the opposite side, the small or grain ends of the straw will then overlap one another in the middle of the mat; all the binders may then be fastened.

Layer after layer of straw is put on this way, the operator working backwards, until the mat is of the desired length. The binders may be lengthened as required, by knotting other strings to them. Care is required to maintain a uniform thickness by putting on the straw in equal quantities, and compressing each layer to the same degree by the binders. When finished, the mat is to be cut from the frame, and the ends securely fastened. The sides are trimmed with a sharp knife, using a straight edged board as a ruler. Mats of this kind, if properly cared for, will last several years. They should, of course, be thoroughly dried before storing away. The engraving given above shows how the successive portions of straw are bound to the warp. (1863)

A CLOTHES DRYER FOR THE FIREPLACE.

Clothes Dryer for Fireplaces

Fireplaces are still used in many houses, and a handy means of drying towels, etc., before them, is a great convenience. Mr. P. C. Waring, Essex Co., Va., sends a sketch of a clothes rack for the fireplace, from which the accompanying engraving is made.

It consists of two light strips fastened by screws to the sides of the mantle, and bearing cross bars to hold the articles to be dried. When not in use, the dryer can be quickly turned up and secured by a small wooden button, and is then entirely out of way. (1881)

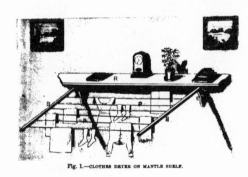

Fig. 1.—CLOTHES DRYER ON MANTLE SHELF.

A Clothes Dryer for a Mantle Shelf

In most houses the stove has taken the place of the open fire, and though the fireplace may be permanently or temporarily closed up, the mantle shelf originally built with it is found quite too convenient to be abolished with the fireplace. Indeed, so desirable is a shelf of this kind, that recent houses built with reference to the uses of stoves only, are almost always furnished with mantle shelves to the chimneys, though there are no fireplaces below.

This shelf, besides being useful to hold lamps and other things, may be converted into a convenient clothes dryer. The great number of portable clothes dryers that have been patented, shows that there is a demand for such things. However objectionable it may be to dry clothes in the house, there will be occasions when it is necessary to do it, and the

simple arrangement here given will answer quite as well as a more expensive patented one.

The dryer in use is shown in fig. 1; *B,B*, are two strips of wood, two or three feet long, as may be desired, and 1½ x 1 inch at one end, and tapering to 1 x ¾ inch at the other. At 4 inches from the shelf, and every 4 inches towards the smaller ends, small holes are bored, through which to pass the lines, *P,P*. The manner of attaching the arms to the shelf, is shown in fig. 2, which represents one

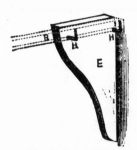

Fig. 2.—BRACKET.

of the brackets into which two small iron hooks, *H,H*, are driven in such a manner that when the arm, *B*, is inserted, it will fit snugly against the hooks, and also against the under side of the shelf. In case the shelf has not brackets, the arms may be supported by means of iron staples, made of the proper shape and size to fit the arm, and driven into the lower side of the mantle. When not in use, the arms can be removed, and the affair rolled up to occupy a very little space. (1876)

Care of Kerosene Lamps

If a list could be presented of the deaths and frightful burnings that have occurred since the introduction of kerosene, it would be appalling. GOOD KEROSENE, that is, of the legal standard of quality, and that sent out by the best makers is far in advance of the legal requirements, *properly* used, need be no more dangerous than the old-fashioned sperm oil, or tallow dips. Filling a lamp while it is lighted is something that ought *never* to be done. It can be avoided by always filling the lamps in the morning. This task should belong to some one member of the household, who should have a fixed and regular time for doing it. It should be made a duty, to be discharged with all the regularity and punctuality of the daily meals. Glass lamps ought *never* to be carried about, for the very reason that they are glass.

In "trimming" the lamps, only the small portion that is charred need be removed from the wick, and that is readily done by scraping with a knife kept for the purpose. If any substance collects upon the wick tube, that should be scraped off, leaving the brass or

metal perfectly clean. After carefully scraping, wipe off the upper part of the wick

tube, (and the wick), with a piece of very soft paper, to remove any small particles left in scraping. A wick may become unfit for use long before it is burned up. Many quarts of oil are carried through a wick, and in time the pores of the fabric become so filled with little particles of dust and other impurities that the oil contains, that its ability to take up the oil as fast as it is burned becomes greatly diminished, and when this occurs, a new wick is needed. If a lamp is filled quite full in a cold room, and then is brought into a warm one, the heat will cause the oil to expand and overflow, and lead to the suspicion that the lamp leaks. This should be avoided by not filling completely; knowing that this may occur sufficient space should be left to allow for the expansion. (1882)

Inexpensive Home-Made Mats

A firm, durable mat, of any size desired, more or less ornamental, and quite inexpensive, may be made as follows: In a capacious bag placed out of sight in a closet, deposit all good remnants from the family sewing, bits of gay colored flannels, cashmeres, pieces of old woollen coats, pants, etc. From time to time, when too weary for other work, and the eyes are too tired to read, cut these materials into strips, as for rag carpet, and sew them together, and wind in balls.

When enough material has accumulated, with large wooden needles, such as are used for knitting shawls, knit these rag strips back and forth into mats of any desired length and width. A variety of effects can be produced by sewing together and winding on separate balls the strips of the same color and knitting them in color bands; or they may be so thoroughly mixed as to give a mottled surface. If the cutting and sewing be done somewhat evenly, the surface of the knitted mat will be quite smooth, and the mat itself firm and lasting. It can be finished with a border of home-made fringe; or with a strip of two shades of cloth cut into scollops, one strip extending a little beyond the other, as seen in the engraving. This is made from a photograph of a mat 15 by 23 inches. (1882)

How to Make a Mud Mat

At any season of the year, mud is one of the housewife's troubles. Those who work in the fields or in the barn yard, cannot help having mud upon their boots, and some means must be provided to prevent it from being carried indoors. Have a few mud mats about the doors and yards, and this trouble will be avoided. They may be made very easily.

Procure a few dozen of common one-inch square fence pickets, three feet or more in length, and a few branches or stems of elder an inch thick. Bore holes through the pickets in four places — near each end and two between — large enough to admit No. 9 fence wire; saw the elder stems into pieces an inch long, and force out the pith; then string the pickets and pieces of elder together alternately. Place washers over the ends of the wires, and after cutting the ends the proper length, rivet them down upon the washers. These mats may be kept outside of the doorstep, and if the boots are rubbed upon them, the soles will be freed from much of the coarser mud or earth adhering to them, and will not muss up the ordinary door mats nearly so much as without their use. (1876)

Spring House Cleaning

Probably there is nothing in the whole routine of house-keeping that is more of a bugbear than the "spring cleaning." If, as is sometimes the case, the house is set in an uproar, the furniture moved out of doors, and the rooms made uninhabitable at once, the operation may well be dreaded. Fortunate are those house-keepers who continue in the old-fashioned method, if colds and sickness among the children do not follow house-cleaning. Of course, where hired help is especially employed for the occasion, it is necessary to keep them occupied and to do in a day as much as possible. But, as a general thing, there is no need of the discomfort that usually accompanies house cleaning. Of course each house-keeper will have her own

views about the matter, and we can only hope to make a few helpful suggestions.

If a regular "scrubber" be employed, she may leave the painted work looking bright and fresh, but it will have been done with soap and sand. Sand should never be used upon painted or varnished work. Strong soft soap and sand vigorously applied will take off a large share of the paint. To clean paint, first make a moderately strong soap suds, and also have at hand another pail of warm water, with a soft flannel cloth for each, and also a plate containing "whiting" or "Spanish white." One flannel being wet with the soap-suds, dip it in the whiting to take up a small quantity, and gently rub the painted work. The surface coating of smoke and other matter will soon be removed, then wipe the surface carefully with the other flannel wrung out from the warm water, and the painted work will look "as good as new."

The chief soiling of wall paper, especially those kinds in which some part of the pattern is slightly raised above the general surface, is due to dust. In many cases all that can be done is, to remove the dust; this is best accomplished by taking a new broom, wrapping a cloth around it, and sweeping with it from the top downward, with long, straight strokes—not up and down. This will remove the dust, and greatly improve the appearance of the paper. Sometimes the paper will be soiled in spots, as where persons have allowed their heads to rest against it. In such cases it is well to try a piece of stale bread, from which the crust has been removed, using upon the spots as if it were a piece of India Rubber.

At house cleaning time it is well to have an eye to, and close up all cracks and crevices, whether in the floor, or between that and the baseboard, where insects may harbor, as well as larger ones in closets through which mice may enter. For cracks, common hard soap, which is usually soft enough for the purpose, may be used to fill them. It is usually soft enough to be pressed with the fingers into crevices, and no insect will venture to make its way through it. For larger holes, through which mice may come, plaster of Paris, mixed with water to the thickness of batter, and quickly applied, will soon set, and stop the opening. Thin sheet-tin, from old fruit cans, may be tacked over the larger holes.

In the spring cleaning we must consider the question of moths. The common clothes moth was formerly the only one that troubled housekeepers, but now there are others. To keep woolens and furs from moths, two things are to be observed—1st, to see that none are in the articles when they are put away, and 2, to put them where the parent moth can not enter. A piece of strong brown paper, with not a hole through which even a large pin can enter is good. Put the articles in a close box, and cover every joint with paper, or resort to whatever will be a complete covering. A wrapper of common cotton cloth, so put around and secured, is often used.

Wherever a knitting needle will pass, the parent moth can enter; carefully exclude the insect, and the articles will be safe. (1882)

Flowers in the House

To renew flowers daily is quite a tax on a small garden, which ought not to be entirely despoiled, for beauty is wanted around as well as in the dwelling. By a little management, cut flowers may be kept fresh several days. The cause of wilting is loss of the water which fills the tissues of the leaves. Evaporation goes on as rapidly as from the growing plant, and there being no roots to supply the loss, the petals and other organs soon shrivel. This is partly remedied by placing the stems in water, but to prolong the period of preservation, it is necessary to hinder the evaporation from the plant.

This can be done very readily by setting the vessel containing the flowers on a plate or pan containing a little water and inverting a bell glass or jar over the bouquet. Water enough should be kept in the plate, to rise above the edge of the bell glass. With this arrangement, as shown in fig. 2, the external

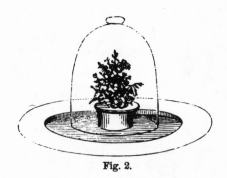

Fig. 2.

air is excluded and the air within the bell glass soon becomes saturated with moisture, so that evaporation from the flowers is prevented and they remain plump and firm until chemical changes cause decay.

Flowers are nowhere more beautiful than when used as decorations for the table. Eating and drinking are purely animal exercises, and the pleasures they bring are lowest in the scale of enjoyments. We should seek to surround this activity with beauty.

Fig. 1.

Flowers mingled with fruits as table decorations are found at all great entertainments. It is pleasing to know that taste in this direction is being encouraged and guided to the right channel. In England the subject has been thought worthy the action of the Royal Horticultural Society. Prizes have been offered and awarded for table decorations. The accompanying engravings, figs. 1,3,4, are specimens of the designs of this kind which took prizes at the June exhibition of the Society. They ranked in the order of the numbers, fig. 1, taking the highest premium. (1859)

Fig. 3.

Fig. 4.

Bouquets & Bouquet Making

To a lover of flowers nothing can be more beautiful than a bunch of them bound together in such a careless way that each flower has sufficient freedom to show its own peculiar habit. We prize such a bouquet, whether culled in our own garden or the gift of a friend, and we place it in water and enjoy its beauty until the last flower fades. This is the bouquet as we find it with those who have flowers in plenty, and among all flower lovers in the country—the real thing.

The city bouquet from the florists is no more like this than a 5th Avenue residence is like a country farmhouse. In one of these natural lovable bouquets there are flowers enough wasted, by being covered and out of

sight, or in their yet undeveloped bulbs, to make a half dozen of the fashionable sort. The city florist, when he sells flowers, is very careful not to sell buds and stems at the same time. In the free way of cutting flowers with their own stems, there is involved a loss of future bloom which he can ill afford, so the flowers are gathered stemless, and the want of their natural stalks supplied by means of art.

With most bouquet makers the next thing in importance to flowers is a broom—a regular corn broom, and this, both brush and handle, is worked up into the most costly bouquets. The broom splints are broken apart and each separate flower is mounted on a broom-corn stem by means of a few turns of a thread-like annealed iron wire as in fig. 1. Sometimes strong elastic grass stems are employed, or pine is split into slivers and used instead of broom-corn, and frequently a coarser wire is used for the stem and is attached to the flower by running it through the lower part of the flower cup, and twisting it below, as seen in fig. 2.

Fig. 1. Fig. 2.
ARTIFICIAL STEMS FOR FLOWERS.

A sufficient number of flowers being prepared in this way, the bouquet is then to be made up, and here is where the broom-handle comes into play. A piece of the stick of convenient length for the center of the bouquet is cut off. As it is customary to use a rose, camellia or other large flower for the center, that is fastened to the end of the stick by running a wire through the flower cup and fastening its ends to the stick by means of small twine as shown in fig. 3. Then the bouquet is gradually built up by adding the other flowers, and securing them in place by turns of small twine. Fig. 3 shows two of the small flowers attached in this way. The shape of the bouquet is governed by desire of the customer, a flat or slightly convex surface in which the flowers are all upon nearly the same level being the commonest form. Any desired shape can be given, and the bouquet will be flat or pyramidal according to the point at which the artificial stems are tied to the central support. All this artificial work is

Fig. 3.—MANNER OF MAKING A BOUQUET.

concealed by an edging of some kind of green.

Flowers treated in this unnatural manner will keep much longer than one would suppose. Of course it is useless to put bouquets of this kind into water, but an occasional sprinkle will keep them fresh for some days, and if they are put under a glass shade, their beauty will be prolonged for a much greater time. (1864)

Inexpensive Household Ornaments

Last month, reference was made to using natural objects for ornamenting rooms. There is hardly a limit to the selection which can be made, for almost everything in Nature is beautiful — tree, rock, flower, moss, bird, insect, and creeping thing — all possess some feature, which the hand of taste can use for its purpose.

This month we illustrated the method of using evergreen cones for making fancy articles, specimens of which were furnished by Miss Anna Pettinger, Kings Co., N.Y. Fig. 1 shows a card basket, made chiefly from the scales of white pine seed cones, and whole seed vessels of hemlock; cones of other evergreens can be used in the same manner, and any desired shape given to the work. The basket here illustrated, was made thus:

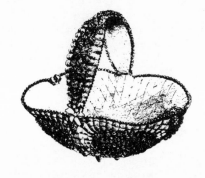

Pieces of pasteboard were cut of the right shape to form the basket, when sewed together. A small wire, properly bent and sewed around the edge of each piece, holds it in shape. Then the more ornamental part of the work as wreaths, rosettes, etc., are made by sewing to the pasteboard small whole cones of hemlock, acorns in their cups, walnut shucks, etc., disposed to suit the taste of the maker. The interstices left are filled with the scales of the pine cones; yellow pine are the neatest. These are sewed on, lapping one over the other, like shingles upon a roof, thus hiding the stitches. A very neat rosette is made by cutting off the bottom of a yellow pine cone. The seed vessels of the tag alder are pretty for trimming. The scales and other parts may be glued or cemented to the pasteboard, but they look neater when sewed. The separate parts when finished, are joined, and the inside of the basket lined with pink or other bright colored silk, quilted in. Open seams left where the edges of the pieces meet may be covered by a beading of allspice berries strung on strong thread or fine wire. Similar trimming is sewed around the edges of the basket, and of the handle. Four large acorns inverted and glued to the bottom, form neat supports.

Fig. 2.

Fig. 2 represents a picture frame, made of the same materials, the pasteboard being fastened to the woodwork; or, the cones might be glued directly upon the wood. For such larger objects, larger cones are appropriate. Those of the white pine split lengthwise with a knife, and the flat side laid down give variety. When all is fastened, the work is covered with varnish, made of an ounce of gum shellac, dissolved in a pint of alcohol. Some give it two coats, adding a little vermilion to the varnish for the first coat, and a very little lampblack for the second, to darken the shade. A finishing coat of copal varnish gives additional luster. (1860)

A Spool What-Not

The old spools that accumulate in a household are sometimes made useful in the construction of a hanging what-not. Three or

SPOOL "WHAT-NOT."

four thin strips of board serve as the shelves. They have a hole bored in each corner for the passage of the supporting cords. The spools are strung upon the cords, and keep the shelf boards at the desired distance from each other. Care should be taken that all the spools used on the four cords between any two shelves are of the same length, or at least are so arranged that the shelves will be level.

"I.J.H.," who sends the sketch from which the accompanying engraving is made, writes that his shelves were made from thin pieces of board upon which dress goods are wrapped. A stain with a coat of varnish will give a pleasing finish to this cheaply made and handy household convenience. (1881)

Window Shelves for Plants

If one has window shelves for plants, it is convenient to have them so arranged that the plants may receive the greatest possible benefit from the light in the daytime, without incurring the risk of freezing at night. To accomplish this, take two of the castings used to hold up the ends of curtain rollers and fasten them on the inside window casing at the desired height. For the shelf, use half-inch stuff, cutting it the shape shown in fig. 2. The width at the widest part may be five or

Fig. 3.—THE SHELF IN POSITION.

Fig. 2.—THE WINDOW SHELF.

six inches; at the ends one inch. Bore gimlet holes lengthwise into the narrow ends as indicated by the dotted lines, slip the shelf between the castings, and put a picture nail through the hole in each casting, and into the gimlet hole in the corresponding end of the shelf. Insert a screw hook in the ceiling above, directly over a point half way between the short edge of the shelf when turned toward the glass, and the same edge when turned toward the apartment. From this hook, No. 9 or 10 wires depend and are hooked upon wood screws in the edge of the short side, the shelf itself being thus supported in a horizontal position, whether turned in or out. If other shelves are required, they may be put between castings fastened lower on the casing, and be supported horizontally by a wire depending from the edge of the upper shelf. The advantage of the arrangement is, that, in addition to the higher temperature secured for the plants by the turning of the shelf, the window curtain can be conveniently interposed between the glass and the shelf at night. If a wider shelf is desired, a suitable block may first be attached to the casing, and the casting screwed to that; or, if the castings are not at hand, a bracket-like block may be used instead. Care should be taken to have the shelf fill the space between the end supports, as in this way it will be stronger, and the whole affair should be as light as is consistent with necessary strength. (1881)

Fig. 1.—THE CHAIR FRAMED.

Barrel Frame Easy Chairs

We have given in past years quite a variety of forms of easy chairs, made from barrels, boxes, hollowed logs, etc. In one of our exchanges from almost exactly the opposite side of the world (the *Illustrated News* of Sydney, Australia), we find engravings of chairs differing a little from any we have given, and we change these somewhat to the forms here shown in figs. 1 and 2. A large barrel or small

Fig. 2.—THE CHAIR COMPLETE.

hogshead with iron hoops, is cut to the form shown in fig. 1, the hoops being first riveted to the staves. Strips or cleats nailed on the inside at any desired height, support the upper barrel head as a seat. The barrel is mounted on a frame of two pieces of wood with casters underneath. A broader firmer base would be formed of three or four pieces. The supporting brackets are added in front. Fig. 2 shows how the whole may be upholstered with calico or any other material at small cost. (1882)

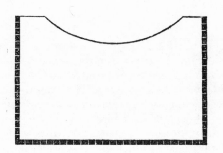

A Cutting or Lap Board

A subscriber, A. M. Ward, Hartford Co., Conn., writes that he was constructing a lap board when the paper containing our description of such a convenience arrived, and thinking it superior, he sends an illustrated description for the benefit of our readers.

The board is 27 inches long, 18 inches wide, and 5/8 inch thick; made of white pine, which should be sand papered smooth, and may be wax polished if desired. Two strips of hard wood are fitted to the ends by by tongue and groove, to prevent warping; this is preferable to the use of cleats. Inch marks are made around the three sides, from left to right, which will be very convenient for measuring any work in progress. Additional finish is given by inlaying a ¾ inch strip of boxwood around the edge, to receive these marks. Both sides of the board are finished alike, though this is not essential. The curve in front to receive the body of the person using it is four inches deep.

Where much work is required, and the board is to be used by a strong person, Mr.

Ward recommends to make it 24 by 36 inches, and the body circle 6 inches deep, and says he prefers to have it square cornered, and without supporting legs. (1863)

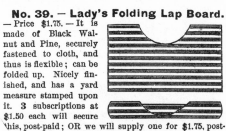
Home-Made Desk

Nearly all persons have papers, letters, and other documents, important and otherwise, which they desire to keep and preserve in good condition. They also need a place in some part of the house where letters can be written and other matters jotted down. For this purpose nothing is equal to a good desk, and to be useful it need not be expensive. Any person at all skillful with tools can construct, in two days' time, a desk similar to the one shown in the engraving, and it will answer the purpose quite well. It is shown so plainly that a short description will answer.

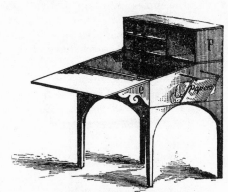

A CONVENIENT WRITING DESK.

The lid is 2 feet 8 inches long and 16 inches wide; when open it rests upon supports, e, that are hinged to the front of the desk, and fall inward out of the way when not in use. The width of the desk is 28 in., and 2 feet 8 in. long. The upper portion, at p, may be firmly attached to the body of the desk or left loose as desired; it is one foot high, ten inches wide, with large and small shelves and pigeon-holes. A row of small pigeon-holes is made in the desk, and should be four inches wide, to readily admit an envelope. It is also best to have one or two small drawers, with keys fitted to them, for the better security of important documents. Papers, magazines, and other printed matter may be placed in the open space in the center. A paper holder is also attached to the side, in which place, papers you have unfinished and other reading matter may be kept. The legs of the

desk may be rounded or left square. The lid when open is, of course, intended to be used as a place for writing. (1882)

A Useful Piece of Furniture

In many rural households, the space allotted to the kitchen is often cramped and narrowed too much. Women are not often consulted when houses are built, and it is usually the kitchen that suffers for lack of room. A piece of kitchen furniture, therefore, that will answer three distinct purposes, is a great convenience.

Fig. 1.—AS A SETTEE.

Here is one, fig. 1, that is at once a settee, a trunk, and an ironing table or bake board. There is a box or trunk, in which one may stow away many things that usually lie about, having no special place allotted for them otherwise. The lid of this trunk forms the seat of the settee. The ends are raised up, forming the arms. The back of it is pivoted upon one side of the ends, and when it is turned down, as seen in fig. 2, it forms a table. When it is turned down, it is held in its place by two small hooks, seen in the illustration at fig. 1. (1874)

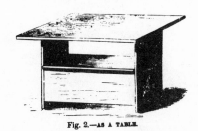

Fig. 2.—AS A TABLE.

Folding Table for Porch

Last summer we found our folding table on the kitchen porch even more convenient than we anticipated when it was first put. Various kinds of work, such as preparing vegetables and fruit, ironing, and the sometimes necessary scouring of tin things, can be done as well on a shady protected porch, as in the kitchen, and with much more comfort, for it affords a pleasant retreat from the heat of the indispensable stove. The table, fig. 1, is made of pine, 3 feet long, by 2½ feet wide. The top is firmly fastened to the wall by two strong hinges. The support in front is nearly a foot

wide, fastened on with a large hinge, so when the table is in use, it is under the top, and even with the front edge. When not in use, the table is hooked up against the house wall, as shown in fig. 2, and is entirely out of the way. Such a table might also be found very convenient in a small kitchen, as it could be let down when any extra work required a second table, and be easily put back when not required. (1882)

Fig. 2.—THE TABLE FOLDED.

A Handy Folding Table

Mr. James H. Ten Eyck, Auburn, N.Y., sends a model of a folding table, from which the accompanying engravings have been made. Concerning the table, Mr. Ten Eyck writes:

Fig. 1.—SHOWING UNDERSIDE OF TABLE.

"It may be made of any desired size, but for the purposes for which it is most generally used, namely, for ladies' work, sewing, as a writing, or invalid table, a very convenient size for the top is two by three and a half feet, and two feet two inches high. Black walnut, ash, and chestnut, are the most suitable kinds of wood. The top should be about 5/8ths inch thick, with bevel or rounded edges; the legs, one by one and 3/4ths inches, rounded edges; the strips or wings at top of legs 1/2 by 3 inches, and three feet two inches long.

The round between the intersection of the legs is about an inch in diameter.

The wings should be hinged to the top, two and one-half inches from the outside edges, and let into the edges of the legs — flush. The spread of the legs on the floor should be about 22 inches for a table two feet wide. Iron or leather washers may be placed between the legs where they cross, to prevent their touching folding. In cutting the rod, or "round," see that it is long enough to crowd the legs apart, so that they will stand a little bracing. This will stiffen the table and make it stand firmly on the floor. This point is

Fig. 2.—TABLE IN POSITION FOR USE.

quite important, and it does not interfere with the folding in the least. To fold the table, simply lift it by the ends. To carry, or move it without folding, lift it by the sides. When folded, it is of a convenient height to carry, and occupies only four or five inches against the side of the room, and is consequently easily set aside, when it is not in use, in an out of the way place." (1880)

Fig. 3.—THE TABLE FOLDED.

Rustic Window Boxes

We have in former years given designs for finely finished window boxes but we have recently seen at the store of B. K. Bliss & Sons some on sale to city customers that we think would suit our rural readers exactly, as they can be made by almost any one.

They are so thoroughly rustic that we have had engravings made of them as a guide to those who wish to try their hands — or to direct somebody else to try his hands — at making them. The foundation in all cases is a box of sound pine, which need not of necessity be planed. The size of the box should have reference to that of the window.

Some windows have sills broad enough to hold the box, but where this is not the case it may rest upon a couple of brackets screwed to the wall.

In fig. 1 cedar sticks, straight and of the same size, are split in halves, the bark left on, and firmly nailed to the box. In fig. 2 is shown what is to our taste an exceedingly beautiful box. It is covered with some well-marked bark; in the case of the one figured that of the whitewood or tulip tree, common throughout all the Western states, is used.

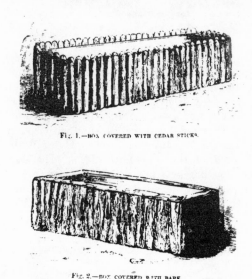

Fig. 1.—BOX COVERED WITH CEDAR STICKS.

Fig. 2.—BOX COVERED WITH BARK.

The engraving shows the manner of laying it on. Fig. 3 shows a more elaborate style, which in reality is more effective than can be shown in the engraving. The ornamentation here is done with halved sticks, those shown light being of white birch, the silvery back of which showed in strong contrast with the darker pieces, which are apparently laurel or some dark-barked wood. In this last case the wood was varnished, which we do not consider an improvement. (1873)

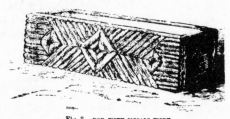

Fig. 3.—BOX WITH MOSAIC WORK.

An Open Wood Box

The engraving illustrates a convenient wood box. The "shelf" is a 1½-inch plank, 1 by 2½ feet. The end pieces are 28 inches high, with the shelf nailed on 10 inches from the floor. A cleat is fastened along the front. The ends may be cut away to form legs, as shown. Such an open box or rack cannot become a receptacle for the refuse commonly thrown into the ordinary wood box. (1886)

A WOOD RACK.

Advertisements.

Advertisements to be sure of insertion must be received at latest by the 15th of the preceding month.

TERMS—(invariably cash before insertion):

HOPPIN.DEL.

Smith thought it would be a fine thing to live in the country. Smith could not get HELP, and as domestic duties began to accumulate and interfere with his case, Smith set his inventive faculties to work, with the above result.—The contrivance is not patented, but is free for the use of all readers of the *Agriculturist*, for whose especial benefit it was sketched and engraved. We can not speak from personal experience of its perfect feasibility.

V Tools

VIEW OF THE TOOL HOUSE OF TOWNSEND SHARPLESS, AT HIS SUMMER RESIDENCE IN BIRMINGHAM TOWNSHIP, CHESTER COUNTY, PENN.

A Tool House—Valuable Suggestions

All ranged in order, and disposed
   with grace,
Shape marked of each, and each one
   in its place;
Nor this along the curious eye to
   please,
But to be found, whene'er required
   with ease.
If used or loaned, and not returned
   by rule,
The vacant shape will show the
   missing tool;
Thus often urged the careless will
   improve,
And rules of order soon will learn
   to love.

The tool house, drawings of which are presented to the readers, is at the summer residence of a citizen of Philadelphia. The building is 20 feet long by 12 feet wide, and is lined with smooth boards. The engravings are exact representations of the building and its interior arrangments, with a few slight exceptions; and notwithstanding there are about 200 tools or implements upon its walls, yet the number may be considerably increased by filling up the vacant spaces with smaller articles, as there may be occasion.

The tools are well secured in their places, and yet may be taken down or put up with ease. They are supported by means of nails, iron hooks of different sizes, stout iron staples, both flat and round, and lighter ones made of wire with the ends sharpened, and of size proportioned to the weight of the tool. The shape of each article is marked upon the wall, with a small stiff brush and ink, and the tools being upon the sides of the building, the floor is left free for other purposes.

Their methodical arrangement, and the shape of each being distinctly marked, combine advantages as to economy of space and security against loss, which could not perhaps be so well attained by another mode, and it is believed to be the secret of causing things to keep in their places.

The writer, with whom the idea of marking out the shape originated, has had the plan in operation for many years, and always with satisfactory results; and the illustrations are presented in the hope they may lead others to adopt this plan. (1862)

A Few Convenient Garden Tools

An old gardener always uses fewer and simpler tools, and does better work, than the novice. We will describe a few of the best, indeed about all that are essential for common operations. A small outlay will purchase the whole of them, and the facility they give to garden work, will pay a very large interest on the cost. Let it be remembered, that a good implement, however high the price, is in the end cheaper than a poor one costing not a fourth as much. Light, well made garden tools, of cast steel, or spring steel, or of iron edged with steel, are always to be preferred to those made wholly of iron.

The spade is necessary to dig holes, drains, etc., to cut turf, to move small quantities of earth in making beds, etc., to divide masses of shrubs or other plants, to take up trees, and the like, but it no longer holds the place of honor.

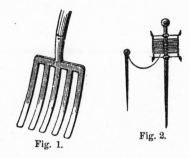

Fig. 1.   Fig. 2.

The spading fork, fig. 1, is the usurper. This is *the* implement for working the soil. It penetrates the ground with greater ease, and

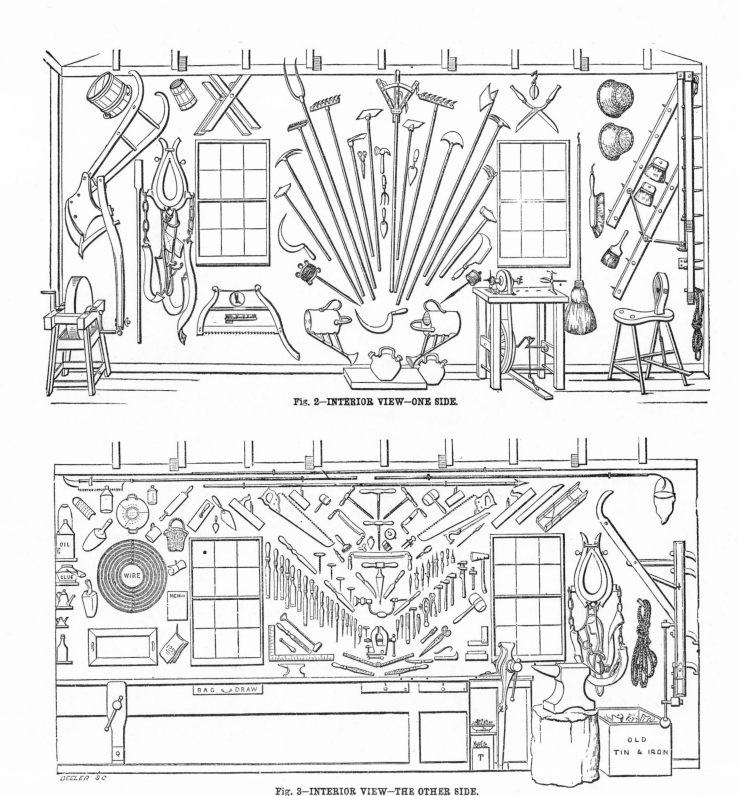

Fig. 2—INTERIOR VIEW—ONE SIDE.

Fig. 3—INTERIOR VIEW—THE OTHER SIDE.

lifts as much soil as the spade, leaving it light and crumbly, not in soggy lumps. We prefer a five tined fork, the tines bevel-backed, and gradually and very slightly increasing in width from the tread to the points, so as to prevent stones from catching between them. For moving loose earth, sand, compost, etc., a shovel is indispensable, except in a very small garden. So also is a common field hoe.

The reel and line, fig. 2, of the form shown in our cut, is most convenient, but any cord wound upon a pointed stake, with another short stake attached to the end, will answer the purpose well. A strong cord of good size is preferable to a string. It should be strong enough to bear a hard pull. The garden line serves a more important purpose in giving a garden symmetry and regularity which marks a well kept place, than any other implement. A long line made so strong as to bear stretching, yet so small in diameter as not to be swayed by the wind, is preferable. It should be housed at night away from dews.

The edging knife, fig. 3. Good grass sods

are far preferable to the box plant as edging for beds. The former, neatly cut and laid, and kept closely trimmed, are neater, more pleasing to the eye, and can be more readily "mended," than the box which often winter kills or fades in spots, and requires a year or two to acquire a respectable size. The grass edging can be in perfection in a month or two. To clip and train the edges of the sod border, a half-moon shaped blade, with a handle like a spade, is convenient, but not indispensable; a sharpened spade is a good substitute.

A steel-toothed rake, the head and teeth

77

being all spring steel, and in one piece, 8 to 12 inches wide, with the teeth long and well annealed, so as not to bend easily, is essential for nice working of the soil, for pulverizing it, for mingling compost with the surface, and for dressing ground walks, removing stones, lumps, etc.

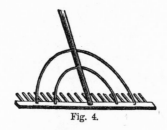

Fig. 4.

A grass rake, fig. 4, is one of the most convenient tools we have used. It is simply a hay rake, with at least twice as many teeth as the common hay rake. The teeth are shorter, and of course closer together. This gathers up the clippings of the grass plot or lawn very clean. We do not make hay in the garden, and the grass never should be more than 4 inches high. All that this rake will not gather is best left on the sward as a mulch for the roots; it will not show, perceptibly.

Fig. 5. Fig. 6.

The shuffle hoe, fig. 5, often wrongly called "scuffle hoe," is a very convenient implement for working among plants. If provided with a long handle, it saves the back from many an ache, and some wear of fingers. The blade should have both edges sharp, and then in shuffling it backward and forward, it cuts both ways, severing the weeds, and leaving them on the surface, and lightening the soil. We advise its general adoption.

Dibbles, fig. 6, of which we present two kinds, are simply round pointed pieces of wood or iron for making holes in which to set out plants or cuttings rapidly. But we do not advise their use in general, for though convenient, a much better hole is made with a trowel or flat stick, inserted and pressed to one side; the soil is not then compressed on all sides. Still the dibbles are very convenient in rapid work, and in the field. The spur on the long one regulates the depth to which it may be thrust.

Fig. 7.

The garden trowel, fig. 7, is very convenient for lifting and transplanting, digging holes, etc. It is like a common small mortar trowel, with the sides curved upwards a little. Our American made garden trowels are less curved than the English manufactured ones, which form nearly or quite half a cylinder; the former are preferable, as they do not cling to the soil, while they answer all purposes.

A knife-blade trowel, or weeder, fig. 8, was recently shown to us by Mr. Theodore Holt, a gardener of New York City, now a missionary horticulturist at Port Royal, S.C. It consists of a blade of steel, an inch and a half wide, and 6½ inches long, tapering with a gradual curve to a point, the shank raised at right angles to the blade, which is sharp at both edges, and nearly flat on the upper side. The use of the implement is in weeding, and thinning out all kinds of vegetables and other plants sowed in rows, and one a little familiar with it, does this tedious work with ease and rapidity.

Fig. 8.

Fig. 9.

The spud, fig. 9, is a stout chisel upon the end of a cane. It should be in the gardener's or master's hand whenever he walks through his grounds; and wherever a weed of any considerable size shows itself, the spud should seek out its root, deep down under the sod or spreading close to the surface, and cutting it off, leave the plant to wither where it stood, or to be easily pulled up. This is good for thistles.

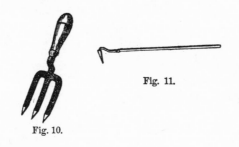

Fig. 10. Fig. 11.

The weeding fork, fig. 10, is a little implement, which we have taken great satisfaction in using of late years. It is the spading fork on a one-hand scale. Penetrating the soil about four inches, it loosens it thoroughly near the roots of plants, and in places where the spading fork can not be used. At the same time it greatly facilitates the uprooting of weeds. For working strawberry beds it is a most excellent implement.

Bayonet or onion hoe, fig. 11. This is a sharp-pointed, double edged steel implement, about 8 inches long and 1¾ inches wide at the broadest part, tapering down to a point. It is set like a common hoe, upon a handle 4

or 4½ feet long. We use it more than any other garden implement. The point, turned to either side, is convenient for working among all kinds of plants, and for digging drills, loosening up the ground etc. The long edge answers the purpose of the common hoe for cutting weeds, loosening the soil without heaping it up, etc. Mr. Holt's trowel, fig. 8, with the shank curved so as to bring the middle of the blade in front of a long handle, and three inches distant, would be still better than the bayonet hoe, we think. (1862)

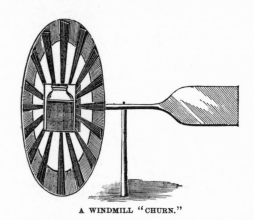

A WINDMILL "CHURN."

Churning With the Wind

Dr. G. P. Hachenberg, Travis Co., Texas, has a "churn" which he describes as follows: "I place the cream in a large glass, air-tight jar, about two-thirds full, and adjust the jar in line of the axis of a windmill, swung low to the ground, as shown in the engraving herewith. The jar is placed inside of a box, which keeps the cream cool, as well as in its proper place. In very hot weather, ice wrapped in woolen cloth may be placed within the box, around the jar. As the windmill revolves, the jar turns with its axis, thus keeping up an agitation of the cream until the churning is effected. A convenient method to churn by this process is to put the cream in the jar in the evening and remove the butter in the morning. The butter is not only made, but thoroughly worked, and taken out in a fine, firm ball. We might call this the tumbling process." (1882)

A Clothes Line Elevator

As usually strung up, the clothes line is almost out of reach, especially at the ends, and clothes are hung upon it with some difficulty, especially by a person of short stature. This difficulty can be quite successfully obviated by the use of the simple arrangement shown in the engraving. The elevator consists of a plank post, *a*, projecting four feet above ground, to which is bolted at the top, near one edge, a lever, *r*, five feet in

of wood, except the crank, axis of the fly-wheel, and pin h.

a, 8 feet long; b, 6 feet; c,c, eight feet; (longer still better;) d, fly wheel of wood, 5 feet diameter, heavy, with crank n, adapted to the depth of the churns e; f, the crank and handle for working-boy or other motive power. The churns should be placed on a board with marks to indicate the exact position of the base of each churn. h, iron pin. Take out the pins, p,p,p, and the churns are detached, and may be removed. (1843)

bottom boards were nailed as shown in the engraving.

It was found convenient for many things to have a box, and one was made of pine boards, which fastened (by means of projecting corner and side cleats) into six holes bored in the frame. A rest-stake — an old broom handle cut down — was attached to the front bar by means of a leather loop, so that it could be thrown over on to the bottom of the cart when not needed to hold the handle from the ground as shown in the engraving.

It is just the thing for carting small quantities of green fodder, grain, etc., for a short distance. (1882)

length. The end of the clothes line is attached two feet from the bolt. The opposite end, three feet in length, is used for a handle or lever for adjusting the clothes line, when filled with clothes, and is retained by a wooden button, b. A small block is nailed upon the post at p, to hold the lever in a horizontal position, while the clothes are being placed upon and removed from the line. A similar "elevator" may be placed at each end of the clothes line, and it may be made of any desired size. (1882)

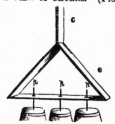

FRONT VIEW OF CHURNS.—(FIG. 54.)

Hand Cart for the Farm

There are many occasions on the farm when a wheelbarrow will not answer the purpose, and a hand cart is just the thing to save hitching up the team. A large cart with good sized wheels is a handy implement.

The one we have used for a long time was made from the hind wheels of an old road wagon. A new axle was made and the irons fitted as carefully to it as if it were a wagon. The tires were re-set and the "running gear" was therefore about as good as new. A body was made out of sound two-inch oak stuff, thoroughly bolted together, and upon it the

A HANDY HAND CART.

Wheelbarrows for Farm & Garden

A good wheelbarrow is a very important implement for both farm and garden, saving cartage, carrying small loads where neither cart nor wagon can go, a great convenience in distributing manure, collecting fruits or crops, moving barrels, and the like. The common form, with flat bottom, movable sideboards and a large wheel, is very convenient for carrying whatever is to be distributed by the shovel, or which should stand upright, as potted plants, or for going through narrow gateways or between close rows, for loading some heavy articles as barrels, and for many other purposes.

Virginia Churning Apparatus

In my late visit to Virginia, I met with an efficient churning apparatus, a side-drawing of which is given below. Its merits are its cheapness and simplicity. A boy or girl of twelve years of age can with great ease work three or four churns. The principle, you will percieve, may be applied to any number of churns, worked by any motive power proportioned to their number and capacity. All the parts are

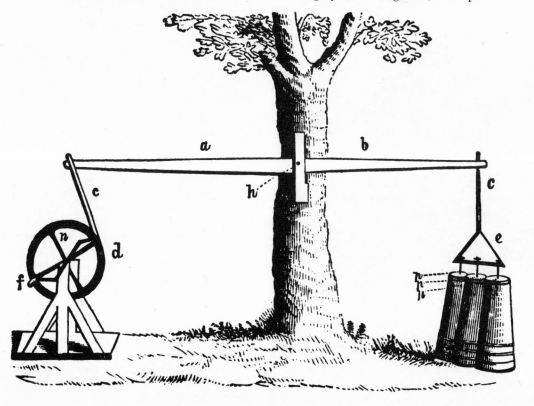

The common "railroad barrow" shown in the annexed drawing is preferable on several accounts. It is much cheaper, costing from $1.75 to $3, according to size and quality. The wheel though smaller is more under the load and thus supports more of it. It may be loaded almost equally well from all sides, the sides being low. The loads are easily "dumped." It may be used for carrying semi-liquid or dripping substances, and is easily arranged to carry grass or hay, by laying in sticks pointing outward on all sides. (1862)

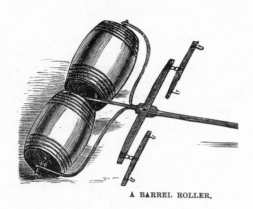

A BARREL ROLLER.

A Barrel Roller

Mr. G. W. Stonecypher, Dawson Co., Neb., sends a sketch of his home-made roller, which he describes as follows: "The roller is made of two coal-oil barrels, filled with soil or sand to give them weight. An iron shaft passes lengthwise through the center of each barrel, upon which they revolve. Two bent bars of iron connect the ends of this shaft with the tongue. Another bar, forked at one end, connects the middle of the shaft with the tongue. The hoops of the barrels should be nailed on; the earth in the barrels needs to be kept moist by occasionally adding a little water. A single barrel may be arranged in much the same way. Such a roller can be used as a 'marker' by fitting circular pieces to the ends of the barrel, or wherever it is desired that the lines be made, placing projecting hoops the proper distance apart." (1882)

The Best Lightning Rod

In general, lightning rods cost a great deal too much, and are often badly made and set up. They are not things to play, or fool with, and a bad rod is worse than none, for it may attract the lightning and then fail to carry it off without damage.

The best rod is of three-quarter inch round iron, drawn to a long, sharp point, which should be made smooth, and gilded, or coppered. The sections must be fastened together with screw ferules, and the ends should be filed smooth and bright, and be fixed in contact.

It is quite safe if fastened to a pole a few feet higher than the building and set near to it, and it should extend ten or twelve feet above the pole. There is no need for glass fittings, as insulators; iron eyes screwed into the pole, or hooks fixed to bands to fit around the pole, are quite as safe as glass holders, which are useless when wet.

The chief point is the ground connection. This should be carried into permanently moist earth, or better still, into water. We might give reasons for all this, but if one wishes to know the way and the wherefore, he should study a handbook of electricity, and he will learn enough to make him quite a match for the peripatetic lightning rod agents, who, as a rule, are not desirable visitors. A perfectly safe and effective rod may be put up in the way above described for ten to twenty-five dollars, and will be worth more than any of the patent fancy things in too common use. (1886)

The "Locomotive Seat"

This is an ingenious contrivance, to save the strain of the backs, and muscles of the legs of persons whose labors require them to maintain a stooping posture, when they have frequently to move short distances, and hence can not take an ordinary stool with them. Especially is this adapted to relieve the nurserymen and gardeners in some of their labors—for instance in grafting and budding near the ground; or setting out plants with which considerable pains have to be taken. The construction is easily seen by the engraving.

The "Locomotive Seat."

An iron sole is firmly attached to the foot; upon this sole and just back of the heel is a socket into which fits a straight ash stick of convenient length, and upon the top of this is a circular disk of wood which affords a very comfortable support to the body, taking the greater part of the weight entirely off the legs. The name "locomotive" indicates that the seat walks with the user. The fact, however, is that the user walks with the seat attached to his foot. It is not in the way of any common movements, and instead of being a temptation to indolence, is rather an inducement for a man to stick to his work, and not find an excuse to get up and walk off somewhere to ease his legs. We have suggested to the inventor its use as a milk-stool, and if the experience of others is like that of the writer, the usefulness of such a stool will be generally recognized. (1864)

A PRIMITIVE PLOW.

A GOOD RABBIT TRAP.

A Rabbit Trap

Mr. S. R. McConn, Bates Co., Mo., sends us a sketch and description of an excellent rabbit trap. He writes: "Rabbits are a great nuisance here, both in the garden and orchard, and a trap of the following kind put in a blackberry patch, or some place where they like to hide, will thin them out wonderfully. A common barrel, with a notch sawed out at top, is set in the ground level with the top. There is an entrance box, four feet long, with side pieces seven inches wide—top and bottom four and a half or five inches. The bottom board is cut in two at b, and is somewhat narrower than in front, that it may tilt easily on a povit at c. A small washer should be placed on each side of the trap at c, that it may not bind in tilting. The distance from b to c should be somewhat longer than from c to d, that the board will fall back in place after being tipped. No bait is required because a rabbit (hare) is always looking for a place of security. The bottom of the box should be even with the top of the ground at the entrance to the top of the barrel. The barrel should be covered closely with a board, as shown in the engraving. Remove the rabbits from the trap as fast as they are caught." (1882)

Crows & Scarecrows

Probably there is no point upon which a gathering of half a dozen farmers will have more positive opinions than as to the relations of the crow to agriculture. It is likely that five of these will regard the bird as totally bad, while the minority of one will claim that he is all good. As usual the truth lies between the extremes.

There is no doubt that the crow loves corn, and knows that at the base of the tender shoot there is a soft, sweet kernel. But the black-coated bird is not altogether a vegetarian. The days in which he can pull young corn are few, but the larger part of the year he is really the friend of the farmer. One of the worst insect pests with which the farmer, fruit grower, or other cultivator has to

contend is the white grub, the larva of the May beetle, June bug, or Dor-bug. It is as well established as any fact can be, that the crow is able to detect this grub while it is at work upon the roots of grass in meadows and lawns, and will find and grub it out. For this service alone, the crow should be everywhere not only spared, but encouraged. We are too apt to judge by appearances; when a crow is seen busy in a field, it is assumed that it is doing mischief, and by a constant warfare against, not only crows, but skunks, owls, and others that are hastily assumed to be wholly bad, the injurious insects, mice, etc., that do the farmer real harm have greatly increased.

Shortly after corn is planted, the crows appear, and are destructive to young corn. Some assert that the crow pulls up the corn plant merely to get at the grub which would destroy it if the bird did not. How true this may be we do not know, but as the corn is destroyed in either case it may be as well to let it go without help from the crow. The first impulse of the farmer, when he finds his corn pulled up, is to shoot the crow. This we protest against, even admitting that the crow does mischief for a short time, it is too useful for the rest of the year to be thus cut down in active life. Let him live for the good he has done and may do. It is vastly better to keep the crows from pulling the young corn, for two or three weeks, and allow them all the rest of the year to destroy bugs and beetles in astonishing numbers.

The corn may be protected by means of "scarecrows" of which there are several very effective kinds. Crows are very keen, and are not easily fooled; they quickly understand the ordinary "dummy," or straw man, which soon fails to be of service in the cornfield. It has no life, no motion, and makes no noise, and the crow soon learns this and comes and sits upon its outstretched arm, or pulls the corn vigorously at its feet.

A dead crow, hung by a swinging cord to a long, slender pole, is recommended as far better than a straw man, as it, in its apparent struggles to get away, appeals impressively to the living crow's sense of caution. But the crow may not be at hand to be thus employed, and if he were, the farmer cannot afford to kill it. Better than a dead crow is a glass bottle with the bottom knocked out, which may be done with an iron rod. The bottle is suspended on an elastic pole by a cord tied around its neck; the end of the cord should extend downward into the bottle, and have a nail fastened to it and within the bottle to serve as a clapper. If a piece of bright tin be attached to the cord extending below the bottomless end of the bottle, all the better. A slight breeze will cause the tin to whirl, and, in the motion, cast bright reflections rapidly in all directions, while the nail keeps up a rattling against the inside of the bottle.

There is also the scarecrow windmill. This is a frame affair on which pieces of tin and highly colored cloth are fastened. Bells are

hung in the upper part of the contraption. (1881)

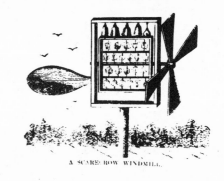

A SCARECROW WINDMILL.

How to Pitch Manure

As pitching manure is laborious work, it is important to render the labor as easy as possible by the exercise of skill in handling the fork, or shovel. To pitch easily, thrust a long-handled fork into the manure, and make a fulcrum of one knee for the handle to rest on. Then a thrust downward with the right arm will detach the forkful from the mass of manure and elevate it from one to two feet high, by the expenditure of little muscular force.

By using a fork like a lever, a man can pitch larger forkfuls, and more of them with far less fatigue, than he can without resting the handle across his knee. When manure is pitched with a short-handled fork, the force required to separate the forkfuls from the mass, as well as for lifting it on the cart, must be applied by the muscles alone. This often renders it fatiguing and back-aching labor. Moreover, when a man pitches with a short-handled fork, he applies his force at a very great disadvantage, as he is required not only to lift the entire forkful with one hand, but to thrust downward with the other one sufficiently hard to balance the force expended in detaching and elevating the forkful of manure. Consequently the arm nearest the manure must expend muscular force sufficient to raise the weight, say, of two forkfuls. This principle is quite as applicable in using the shovel as the fork. By resting the long-handle across one knee when shoveling, keeping the arms stiff, the body erect and straight, a slight thrust of the body and knee will force the shovel into the earth with the expenditure of little force. (1865)

VI Building

Fig. 1.—PERSPECTIVE VIEW.

Pioneer's House—Costing $250 to $500.

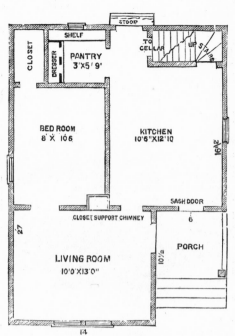

Fig. 2.—FIRST FLOOR.

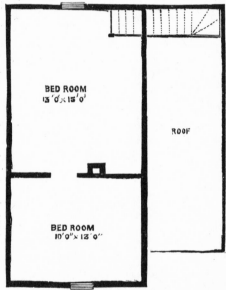

Fig. 3.—SECOND FLOOR.

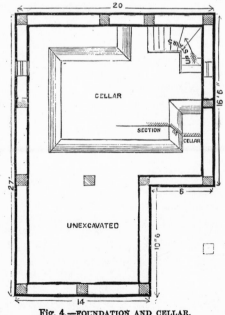

Fig. 4.—FOUNDATION AND CELLAR.

Choosing a Building Site

One thing to be considered in choosing a site is its accessibility. Persons of a poetical or romantic turn of mind might choose to live perched on a high and wooded hill, remote from the common thoroughfares of men. But sober-minded people would inquire, how are we to get to the house in dark nights, in muddy weather, and in winter?

The surroundings of a site should be considered. Who, unless he be the smithy himself, owner and proprietor, would want to build a house with the smoke and dirt of a forge just under his windows? Cattle sheds, pig-stys, a slovenly neighbor's out-houses, any kind of building used for manufacturing purposes, a slaughter-house, a quagmire — in short, any thing and every thing in the line of nuisances ought to be avoided. Whoever builds in the neighborhood of a nuisance will find it a source of continual mortification and regret, and will be quite sure, ere long, to sell out his property at considerable sacrifice.

The healthfulness of a site is an important matter. One would hardly wish to build on the edge of a marsh, or of any standing water, unless he is proof against malaria. Low grounds generally are subject to dampness in summer, and to extreme cold in winter. But in avoiding low grounds, let us not go to the other extreme of pitching our tent on a high and bleak hill. A site moderately raised above low flats, yet not on the hill top, is ordinarily the most salubrious, as it is the most pleasant. Such sites afford pure air, pure water, good drainage, and cheerful prospects, all of which are conducive to health and happiness.

The nature of the soil should be looked to. Whoever builds in the country, builds there principally for the sake of having a little farm or garden and orchard, and ornamental trees. But these will not succeed well on a rocky bluff, a sandy plain, or a bed of clay.

Any improvements already made upon a site should have considerable weight in determining one's choice. An orchard, not too old and scraggy, a few good shade trees of large size, any amount of grading, draining, manuring, fencing, etc., that may have been done, is only so much time saved, and so much money invested for the buyer's benefit. Pleasant as it is to create one's homestead wholly, yet life is short, and it is quite a help to have a few things already created to one's hand. (1862)

84

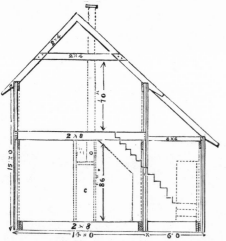

Fig. 5.—SECTION OF FRAME, ETC.

Fig. 1.

A Laborer's Cottage

We give the simplest plan of a farm cottage for a working man and his family—the latter not large, of course. It is, in the main part, 22 by 14 feet, with a lean-to 8 feet wide on the rear, and projecting 4 feet at one end; all covered under one roof. The elevation, from the bottom of the sill, to the top of the plate, is 12 feet, in the main part, and several feet on the rear of the lean-to behind. The roof is a "quarter pitch," or 3½ feet, being one foot perpendicular rise to four in the width of building. This pitch is sufficient to give a good flow of water down the roof, but it may be increased to one-third, or a rise of one foot to three in width, if higher chamber room be necessary. The roof is a hanging one—that is, it projects 18 or 20 inches over the walls of the house, so as to thoroughly protect and shelter them from storms and weather, besides adding greatly to the warmth, comfort and appearance of the tenement, and carrying the water completely off from all drips along the walls, and throwing it to a distance from the sills and underpinning. Just above the edges of the roof, also, can be placed gutters to carry the water to one end of the house and throw it into a cistern, if necessary. This style of roof, having no breaks, or angles in it, is lasting, and if well laid, leak proof—which if broken by modern hips and angles, it would not be.

The sills are 8 inches square, and the joists 3 by 4 inch common scantling, laid crosswise, and 14 feet long; or, they may be 2 by 6 inches laid edgewise, and of either size, not more than 2 feet apart. The lower rooms are 8 feet between joints, leaving a chamber of 2½ to 3 feet perpendicular wall below the plates, and the height of the roof above—sufficient for such a tenement. The lean-to rear is but a shed of course, without chamber floor, the roof running continuously down in salt-box fashion, with long rafters over its back wall. This is shown in fig. 2, as the end elevation is not shown in the perspective. Yet,

it is covered in with the same material, and in the same manner as the body of the house. It is a "plank" house; that is, built of inch or 1½ inch 12 feet unplaned boards or plank, placed up and down perpendicularly, and battened over the cracks with 3 inch strips of inch boards; and if a better finish than this be required, matched boards or planks may be substituted, with battens over the joints, or clapboards, either planed or rough, can be laid over them; and if the covering be planed, painted; if not, whitewashed.

INSIDE ACCOMMODATION.

Fig. 3—GROUND PLAN.

The front door opens into a living room 14 feet square, in which is a stove for cooking, and the pipe runs up through the chamber floor into the chimney, standing either on the floor, or hung by a gallows suspended from the rafters. If the latter, the pipe goes into it through a thimble or crock inserted in the chimney by an elbow from the tops. At the left of the entrance door is a bedroom 8 feet square, which may be entirely separated by a close partition and door, or may be an alcove simply, separated by a curtain, so as to be, in reality, only a recess for a bed, table, and glass. Next this bedroom is a flight of stairs 3 feet wide, leading to the chamber. Next to, and under the stairs, is the family "buttery,"

or provision, and dish closet, 8 by 3½ feet, with a window, and shelves, as convenience may demand. A door leads back into the lean-to, or shed, in which may be an outer cook room for summer, a bedroom, or whatever else is required, and a woodhouse, with a door leading outside at one end, as shown in the plate. We have not partitioned this in the plan, leaving it for the builder to appropriate as may seem best.

The windows of the cottage are "hooded," that is, sheltered by short strips of board 10 or 12 inches wide, sloping outward, supported by brackets at the end to ward off the violence of storms, and keep the upper joints dry, besides giving a sheltered, cosy look to them. So also, is the outer lean-to door hooded in the same way, only that its hood is 3 or 4 feet wide, and the brackets in proportion. A small stoop or verandah is thrown over the front door, say 6 feet wide and 4 feet deep, with a seat on each side. All these outer appendages give the cottage an air of completeness, and repose, which, aided by a few climbing vines or shrubs, make it all that can be desired as a rural cottage.

We will add, that if the cooking stove be removed into the lean-to for summer use, the pipe can run out through the roof above by displacing a few shingles, and inserting under

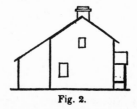

Fig. 2.

the ones that are above it, a zinc plate, through which the plate passes—a lip being turned up all round, except on the lower edge, to carry off the water; and when the stove and pipe are removed, the plate can be taken out, and another whole plate put in, or the shingles replaced, so that no leakage can occur. (1858)

An Extended Farm House

The original of this, in its cottage size, was 26 x 18 feet, with a lean-to 14 feet wide, and an extension, as now shown, of 16 feet in length on the main part, and on the lean-to, of 20 feet, and an upright height of 16, instead of 12 feet, with the roof running down over it, at a quarter pitch. The height of the lower rooms, in the main body, is 9 feet, and of the lean-to rooms, the same; or they may be reduced to 8 feet; and in case the pitch of the roof will not admit of that high in the rear lean-to posts, the ceiling over head in the rear can follow the rafters until the proper height is reached, and then pass on a level to the partition, making a good finish, and sufficiently well-looking for the humble purposes to which these rear rooms were used.

On the front is a verandah 40 x 8 feet, the whole upright part of the house being 49 feet long. The entrance hall is 9 x 6 feet, leading on the left to a commodious parlor 18 feet square; on the right is a smaller sitting room 18 x 13 feet. Between these rooms is a chimney with two flues, admitting a stove pipe from each, and as a closet for each one is between. In rear of the parlor is a family bedroom 14 x 12 feet, adjoining the inner or winter kitchen, which is 20 x 14 feet, having a closet next the parlor of 6 x 4 feet. Between the sitting room and inner kitchen is the stairway, commencing to lead up at the inner end, with a cellar door, and stairs to go down, at the outer, or right hand extremity. Next the kitchen, on the right, is a buttery, or provision and dish closet, 12 x 10 feet. A passage of 4 feet wide between this closet and the stairway leads into an other, or summer kitchen, 18 x 14 feet, entered in front from the verandah, which is 19 x 6 feet.

In the inner kitchen on the rear, between the windows, is a chimney with stove flue, fireplace, and oven, if all three of them are wanted. This chimney may be dispensed with, which we would prefer, and a main chimney built where the 6 x 4 closet stands, having a broad, old-fashioned fireplace, and oven in it, and connected with the other front room flues above, so as to have but a single chimney stack through the ridge of the roof, as in fig. 4. We do love an ample, open fireplace in the farm house kitchen, especially in a wooded country; and if coal be used, we equally like an open fireplace and grate. It is a great, and most genial

dispenser of warmth, and comfort, and nothing so much promotes good fellowship in winter as a cheerful fireside. At the end next to the woodhouse is a chimney, with arch, and kettle, a pipe flue, or more, a fireplace, if needed, or any other heating use required. Adjoining it, runs off to the rear, a wood house with half its gable end projecting to the right, forming an open en-

building needed.

We have thrown the broad, or hanging roof over this house, and would have it project over the walls of the main body 24 to 30 inches, and 18 or 20 inches over the wing, and woodhouse. We have also given it a full front verandah in both house and wing. We believe in verandahs — as much so as in any room inside of the house itself.

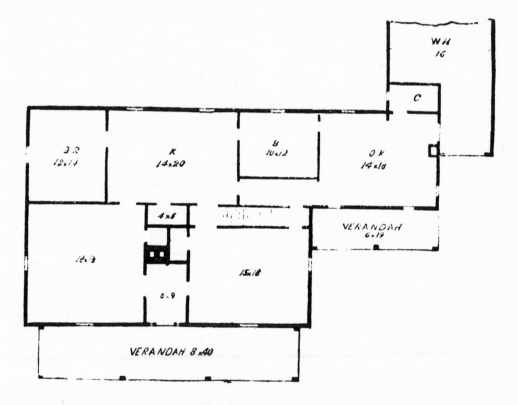

Fig. 5—GROUND PLAN.

trance. Immediately in rear of the kitchen is a closet, or summer buttery for milk, or other purposes, partitioned off from a part of the woodhouse. This woodhouse is 16 feet in width, and may be extended to any depth back required, and may there join into a workshop, carriage house, or any other

In the first place, they don't cost much; secondly, they afford shelter, shade, and protection; and thirdly, they are the very pleasantest parts of the dwelling to sit in during the warm season of the year, and enjoy the social intercourse of family and friends. (1859)

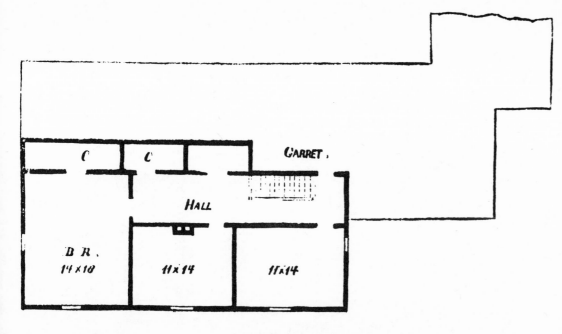

Fig. 6—CHAMBER PLAN.

How to Build a Log House

There are two kinds of log houses — one, the unadulterated rough, round-log tenement; the other, the logs hewed down on two sides, set edgewise each upon the other, and called by distinction, the block house. This is, ordinarily, the second degree in luxury from the primitive habitation of the first backwoods settler. We have had difers experiences in each of these descriptions of house building; accounted ourselves a master workman, even among the craft, and after five and twenty years interregnum in that necessary branch of architecture, firmly believe that we can yet "carry up a corner" equal to the best of their builders.

To commence: "The proprietor" selects his site, cuts down the heavy trees within "falling" distance of the spot the future house is to occupy, and clears away the stumps and underbrush close to the ground. The day fixed for "the rolling," his neighbors — to the number of ten, a dozen, or twenty, according to the magnitude of his building, and the extent of the finish to be given to it — are invited; and after an early breakfast, with two or three yoke of oxen or spans of horses, as they may own them, assemble on the ground for action. The company are then called together, and some one, usually conceded by the company to understand the matter thoroughly, is agreed upon as "boss" for the day. Four athletic, active choppers, each with a true eye in his head, are then selected as "corner men." There are more, if the house is to have log partitions — one to each "butt" or at the intersecting point where the end of the transverse logs lie upon the bodies of

the front and rear of the main ones of the house, as in fig. 2.

A man with an ax stands upon each intersecting point to "carry up the corner," the foundation being first laid by a course of heavy logs of durable timber laid flat on the ground, and on them the sleepers for the main floor; or, if convenient to be had, a large flat stone is laid under each corner of the building, and, if a double house, under each partition. This preliminary labor is sometimes done a day or two previous, by the owner and the help of a neighbor or two, but in most cases is left for the day when the rolling up of the house itself is to be performed.

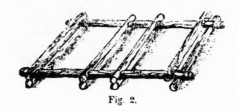

Fig. 2.

Well, the company assembled — the men partitioned out, each to his allotted branch of labor, into "boss," corner men, teamsters, and choppers — they commence work in earnest. The boss selects the trees, the choppers fell and "butt" them at the lengths which the boss marks them out; for he, with ax and pole in hand, must be round among them constantly, to see that no mistakes are made in these particulars. Or, sometimes, when a very rough house only is intended, each chopper selects his own trees, and draws his own measurement by the aid of his ax-helve, or "paces" it off, and even lets the butting go; but the measurement and butting is the better way,

usually. As fast as the trees are prepared, the teamster, with handspike in hand, is ready to hitch on to one end, and his cattle, with a wonderful knowledge of the kind of work from being used to it, are off in a moment; and, after one or two logs are hauled, they stop with great accuracy at about the very point where the logs are required for rolling. It is also well understood by the teamsters — the word being given by the corner men as they proceed — whether the but, or top end of the log is to go forward, for the house must go up on as near a level as possible. This is accomplished by putting the but end of a log at the corner which happens to get the lowest.

The log being at its place on the ground, four or more men, as may be required, with handspikes (levers), and their stout arms and shoulders, roll it up on skids (pieces of timber with one end on the ground, and the other on the log last put up). When breast high, it is carried further up by aid of crotched poles cut from forked saplings. When nearly up, to get it over the projecting cor-

Fig. 3.

ners, the corner men, with handspikes or larger levers placed across the last laid log, raise it over the projecting ends of the transverse log. They lay it on its proper side for "notching" to fit the bearing logs, which have been previously "saddled" — that is, scarfed down from the top on each side, like this letter   (b, fig. 3). They then cut into the last received log, a corresponding notch to fit closely upon the other, thus V (a, fig. 3); then roll it over, and if properly done, it fits snugly, and, with the aid of a correct eye, the corner is carried up perpendicularly from sill to plate. In the process of going up, the places for the doors and windows are either marked on the logs by the corner men by cutting a scarf into them, or cut out altogether, as they may have time. When the first, or lower story is up to a sufficient height, the beams are laid on, scarfed in by notching and saddling, according to the extent of finish to be given to the house. The bearers, however, are previously flattened, so that their upper sides may receive the floor when laid upon them. Thus the house goes on until the proper height is reached; and, if a very rude one, the gables are then laid up in the same manner, only that the gable logs are scarfed off at the ends with a slant to give the roof the desired pitch. The rafter logs, laid lengthwise, are notched upon them so that when finished with its ridge pole, it is the perfect skeleton of a future cabin, in all its majestic length, breadth, height, and proportions.

Thus, the rolling is finished, and according to its magnitude and extent of hands employed, may take only a forenoon, or the entire day. This heavy chopping and lifting, if at a regular "Raising," is seldom done without at least something good to eat, and it used to be something to drink about once in an hour or two — a custom happily gone out of fashion, in part at least. Then comes a vigorous and hearty dinner of baked pork and beans, and other substantial nutriment, at "noon-time," eaten on a clean basswood, maple, or hickory chip, with jack-knife for carver, and fingers for forks, a merry crack of jokes, and a generous "nooning" afterwards.

The next day, or soon afterward, the owner returns with an extra man or two to assist him, and in the course of a day or two finishes up the house, by cutting out doors and windows, laying the floors, and putting on the roof. In the absence of boards, the doors and floors are made of "puncheons," that is, logs split into short planks. The roof is covered either with boards or newly peeled bark, laid lengthwise from the ridge pole to the eaves, and battened, to keep out rain and snow; or, more frequently, perhaps the roof is covered with thin staves split from oak, laid on and held fast by poles which are withed at the ends, to keep them in place and firmly pressed upon the staves to make the joints close. If there be time, the inside logs are hewed down to a face, the corner men, or an extra hand or two having "scored" them with axes as they were rolled up. A clay hearth is laid at one end, a chimney, built up with split wood and clay, and the house is ready to move into, in a day or two, at farthest. Such is the history of a pioneer log house in America. (1857)

Smoky Chimneys

Chimneys on the one-story wing of houses are often caused to smoke by the wind blowing over the top of the higher part of the house, and down into them. A multitude of contrivances have been devised to remedy this evil.

The most common is a cap of stone or iron laid upon two courses of brick at the four corners of the chimney, thus: as shown in fig. 1. This answers pretty well when the wind blows down steadily from over the house, or from any direction except towards the house. Then by striking against the side of the upright part of the house, it is broken into a thousand eddies, driving this way and that, up and down,

Fig. 1.—SIDE ELEVATION OF COTTAGE.

A Country House, Costing $500 to $800.

under the chimney cap and down the flue, and filling the house with smoke.

Several devices are in vogue for meeting this difficulty. One, highly recommended, is called Mott's Ventilator, of which fig. 2 is a sketch. We have seen a modification of this in a cheaper form, which can be made by any worker in tin and sheet iron. Get an upright piece of stove pipe, two feet long and eight or ten inches diameter, and make it square at the bottom so as to fit the flue. Fasten this to the top of the offending chimney, by brick work or by a cap of sheet iron, the first is preferable. On the top of this upright piece, fasten a horizontal section of the same diameter, but flaring a little at each end. It will look somewhat like fig. 3. We have known this simple contrivance to work well on chimneys where several fashionable and costly ventilators had failed to afford any benefit. (1859)

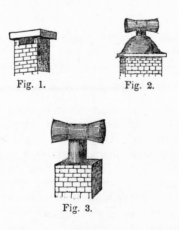

Fig. 1.   Fig. 2.

Fig. 3.

Painting Houses

In going through the country, the eye is wearied by the steady succession of white houses, usually with green blinds and red chimneys. Why we see so few houses of other colors I am unable to say. Perhaps white is adhered to from force of habit. A house so dazzling in its whiteness that it could be seen from anywhere within a goodly circle of miles was our fathers' highest idea of beauty. I can conceive of no more self-asserting and disagreeable feature in a landscape than a great white house, standing in an open yard, unless it is a red one. In summer the contrast between a white house and the landscape is too strong. In winter there is none.

Before painting a house, we should study the landscape about it, and decide on a color that will be in harmony with it. We do not want a green house. I do not mean that when I say we should select a color in harmony, but I mean a color that is in contrast with the prevailing tints of the landscape, and does not conflict with them.

For country houses I would advise for open, exposed places, a pale gray, or drab. There are complaints made frequently that drab looks cold. It can not look colder than white does, and there is no reason why it should look cold at all, if proper care is taken to have the trimmings of the house of some warm, cheerful color. I know a drab house with deep, warm-toned brown cornice and blinds, with plenty of vines clambering up it to break the monotony of the surface between the windows, and it is one of the warmest-looking houses I know of. The general effect in summer is cool and subdued, and in winter it gives a sense of warmth and comfort. (1882)

Fig. 2.—FRONT ELEVATION OF COTTAGE.

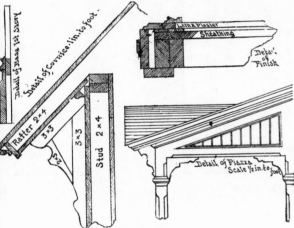

Fig. 6.—CORNICES, FRAMING, ETC.

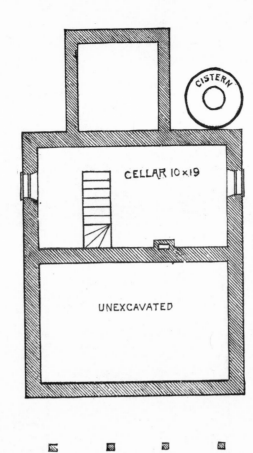

Fig. 3.—PLAN OF CELLAR.

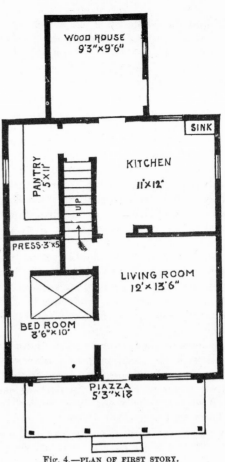

Fig. 4.—PLAN OF FIRST STORY.

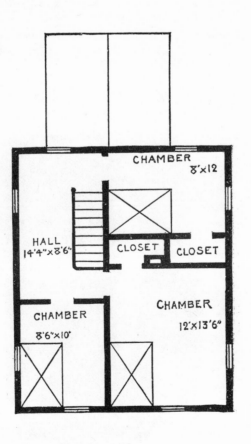

Fig. 5.—PLAN OF SECOND STORY.

Cheap Paint for Rough Fences, Outhouses, etc.

The *Scientific American* says: "Pulverized charcoal and litharge (oxide of lead) in equal quantities, mixed with raw linseed oil, makes a cheap and very durable dark-brown paint for rough boards of outhouses. It is also good paint for exposed iron work. The addition of yellow ochre makes it a dark green color."

This appears to be a good recipe, if charcoal can be conveniently pulverized finely. It would need to be passed through a fine sieve to remove the lumps of coal. Probably the coal dust with the litharge would form a good body, while the litharge would make it dry rapidly. For producing black color merely, it would be cheaper to buy lampblack, which is in reality a very finely divided charcoal collected from the smoke of resin or turpentine. In ordinary times lampblack is very cheap and might be used instead of the charcoal. The greenish tint given by the yellow ochre would be preferable to the dark or blackish brown of only the litharge and coal or lampblack. (1862)

A Ground-Level Barn

We have studied the plans of a great many barns for ordinary farm use, theoretically as well as practically, and taken altogether, the one we now present is as good as any, if not the best that we have seen. As we understand it, when the whole of its accomodations are not required, this barn may be built, in a part of its arrangements, by sections, with equal facility and convenience as the entire structure; and the remainder added at a future time, if needed.

This barn is a large one — much larger than is usually required on small farms; 100 x 50 feet on the ground. It is placed four feet above the surface on wooden posts of a durable kind, or on stone piers, or on a continuous wall of underpinning. The posts are 18 feet high, from sill to plate, with a roof of 17 feet, or one-third pitch, to give the water a rapid passage off, and admit more storage underneath. In the center of the roof is a ventilator, to pass off the moisture inside, if any. An inclined plane of plank, or earth, at each end, leads to a floor, which is 14 feet wide through its whole length.

The bays on each side are 18 x 70 feet. This allows 25 feet (the width of two bents of 12½ feet each) at the front entrance, on each side, for machinery, threshing floor, &c., with a flight of steps on the right hand side to the granary overhead.

Around the whole building, 3 feet below the barn sills, and 1 foot above the ground, is a line of stables, 16 feet wide, with the outside posts 12 feet high, tied by beams, 7 feet from the floor, into the main posts of the barn. The upper ends of the stable rafters rest on a line of girts between the barn posts, one foot below the barn plates. These stable roofs have an 8-feet, or one-quarter pitch, and under them, on a scaffold, is large storage for hay, or other cattle fodder. The stables have passages of four feet next the bays, which connect with the barn floor by a passage of five feet at the further end, to receive the hay thrown down from the rear side of the bays. The mangers are 2½ feet wide. The stalls are double, or for the accomodation of two animals each. The end stables can be devoted to miscellaneous stock uses, as may be convenient. The doors will be seen by the open spaces on the floor plan.

The lines of sheds may be either built or not; though for stock purposes, they may be considered indispensable; they may be omitted, either in whole, or in part, or extended as the wants of the farm require. In this plan they are thrown out 64 feet, at right angles to the barn, in the rear, and extend 116 feet on a parallel with its sides. They are 16 feet in width — of the same height with the stables. They may be partitioned partially into stables, as in the plan, or used without. The barn has two large cisterns to hold the water from the roof, to accomodate the stock. (1858)

Pennsylvania "Double Decked" Barn

The barns in different sections of the country vary in many points, and could plans of what in each section are considered models of excellence be presented, it would be very instructive. We anticipate being able to publish several good plans from distance localities, and this month give one from Chester Co., Pennsylvania, furnished by Alfred D. Sharples, who writes thus: "I have been eating the good fruit of your table for the last seven years, and it occurred to me quite recently that I ought to help replenish in equal ratio. I send a design for what Pennsylvanians call a 'Double Decked Barn.' It is built on a hillside, to face north or east. This one, suitable for a farm of 100 to 150 acres is 64 x 60 feet; ceilings of the lower stories 10 feet high; 3rd floor 16 ft. to the eaves. I believe this building combines the advantages of all the tumble-down concerns usually found on a farm, in as neat and compact a form as they can be placed."

Friend Sharples does not mention to what use he puts the space under the corn cribs, entered from the middle floor. So we will call

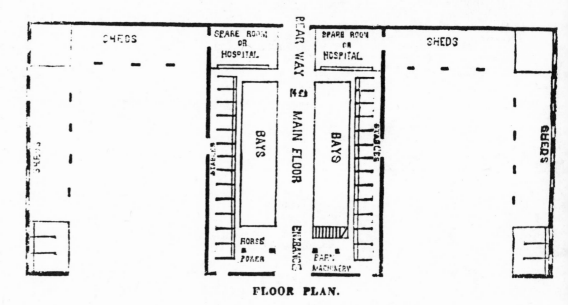

FLOOR PLAN.

it a fruit room, with an ice-house in the rear of it—that is, under the slope, or "proach" as it is sometimes called. The ventilators, used also as hay shutes, appear to discharge into the peak of the roof. This is undesirable and one or two outside ventilators, like the one represented, fig. 5, near to which the ventilating trunks should rise, would make the ventilation more effective, and the breath, and exhalations of the stock would not be condensed upon the hay in the loft.

Fig. 5.—VENTILATOR.

Scarcely two farms, or two farmers, rather, have the same wants, and so any general barn plan must have considerable elasticity, that it may be adapted to these varying necessities. When, for example, there is no good side hill whereon to place the barn—the middle floor might be entirely omitted, some of the cattle stalls put in the shed, closed in, and thus space made for the grain bins, tools, etc., on the ground floor. The amount of corn raised upon different farms varies so much that in many cases such provision as here made would not be adequate. The loft above the shed, or an independent corn-crib would be necessary. Hog pens are perhaps kept away from the barns in Chester County. This is not a bad plan in some respects, for where most economically raised and fattened swine should have cooked food, and fire close to a wooden barn is undesirable. Still at the end of a 60-foot L a boiling house with hog pens adjacent would be conveniently near, and yet distant enough to avoid danger from sparks, if coal fuel is used.

It will be noticed that the threshing floor is nearly 18 feet wide. This is much wider than

REFERENCES.—I, Carriage room—J, Tool room—K, K, Hay—L, Coarse Fodder—M, Straw—N, Cribs,—O, Passage—B, Grain Bins—Q, Work shop—S, S, stairs—V,V,V,V, Ventilators.

Fig. 3.—FIRST DECK.

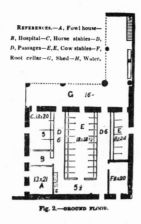

REFERENCES.—A, Fowl house—B, Hospital—C, Horse stables—D, D, Passages—E, E, Cow stables—F, Root cellar—G, Shed—H, Water.

Fig. 2.—GROUND FLOOR.

is generally considered necessary; 15 feet is wide enough. This width of the floor would admit of larger hay and straw bays, of larger corn-bins, and besides, of the great convenience of sliding doors. Sliding doors about a barn, particularly great doors arranged on top rollers, are among the greatest improvements of modern farm architecture. In high winds swinging doors are really very dangerous—and in such a position as the plan presents, the great doors are particularly exposed and might easily cause fatal accidents. Sheep raisers will look in vain for a place for their favorites. The sheep quarters might be provided in sheds, extended more or less, according to the size of the flock—or a rearrangement, to a certain extent, of the ground floor might be made for their accomodation. (1864)

Rustic Work Structures

The term "rustic work" is now used for many objects made of materials, the surface or the shape of which is left in the natural condition. The smallest flower baskets, consisting of a bowl ornamented with cones and crooked sticks, and large, even elegant edifices, such as are seen upon our parks, are classed under the rather comprehensive name of rustic work.

Probably no finer specimens of this style of architecture can be found anywhere than at New York Central Park; the shelters, summer houses, seats, arbors, boat landings, and bridges, built in this manner, are numerous, and are tasteful in design and executed in a workmanlike manner. It is probable that the successful introduction of rustic work at the park has done much towards popularizing it, for we now seldom visit a neighborhood where any attention is given to rural adornment that we do not see more or less ambitious attempts at this kind of decoration and frequently excellent examples.

Work of this kind should present the expression of durability and solidity. Its very rudeness of exterior demands that there should be nothing shaky about the structure. There is no wood so well suited to the purpose as the red cedar, not only on account of its great durability, but because the natural growth of its branches presents a great diversity of angles and curves, twists and knots, that in the hands of a skillful workman give most pleasing effects; besides these, its color is a harmonious one.

No instructions can make one a clever builder of rustic work, he must have a natural ingenuity that will allow him to combine irregular shapes into something like symmetrical forms. A mere association of grotesque branches is not pleasing. There must be an architectural design, and the details of this worked out by the ingenious use of natural materials. We give a few illustrations of simple structures. In fig. 1, we have a bird house and a support for climbers combined. The central pillar is made sufficiently strong to support the structure, and the vines are trained to the corners by means of wires. Fig. 2 is a bridge upon the estate of Edwin A. Saxton, Esq., at Tenafly, N.J. Rustic work is often

Fig. 1.—BIRD HOUSE.

Fig. 2.—RUSTIC BRIDGE.

used with fine effect in small bridges, and though this is less regular in its design than some we have seen, the effect is very pleasing. The covered arbor, fig. 3, is an exceedingly simple design. It is over one of the path ways at Central Park. Fig. 4 is a

bee stand at Central Park. The roof, and enclosed sides, and ends are covered with split sticks of red cedar. (1870)

Rustic Summer House

The violent contrast between the irregular yet graceful forms of trees and shrubbery, and the angular and precise finish of dwellings may be in some measure subdued

by introducing into the house surroundings a rustic style of architecture, combining natural and artificial features. The design would be quite appropriate for some half-secluded nook in a large landscape garden, when it would add an attractive feature to the adjoining lawn. Something less pretending and elaborate would answer better for grounds embracing not more than two or three acres. The general style, however, might be preserved.

The structure here illustrated is somewhat large and elaborate. But simple, cheap ornaments may be readily provided for the plainest and most lowly country or village home. We have seen a pleasant summer retreat constructed with a few cedar poles, some set up for posts, and others fastened across with withes, or nails, and twigs were woven over the top and part way down the sides. Morning Glories were planted around the border, and trailed on the whole. In another case, the covering was made with larger twigs and ivy trained up the sides and over the roof. This looked quite pretty in winter. (1859)

A Summer House Grape Arbor

In one of our recent excursions, we saw a very simple, easily made, cheap, and yet pretty structure, which answered the double purpose of a grape arbor, and a tight roofed summer house. The accompanying engraving is as correct a representation as we can give. (This structure is at the residence of Mr. Mumma, just west of Mechanicsburg, Cumberland Co., Pa.)

Four posts are set up about 10 feet apart, to form the corners; these are of undressed cedar, if we remember rightly. At the height of eight or nine feet, small timbers for plates are spiked on, and a four-sided roof, somewhat flat, runs to the center. Lattice work occupies about one-third of the space between the corners and also around the top of the two arched entrances on each of the four sides. The four corner posts are carried up five or six feet above the plates, and horizontal strips are nailed from corner to corner. The top edge of the upper strip is made wider and cut a little ornamental. From the peak of the roof, in the center, a square shaft rises a few feet, terminating in four points, between which stands a weathercock. A grape vine upon each of the fours sides runs up along the lattice work, and spreads out upon the horizontal strips. These were well loaded with fruit when we saw them. The floor consisted of a bed of dry, spent tan bark.

The whole structure has a light, airy appearance, and when adorned with the fruitful vines is certainly ornamental. No great amount of labor or expense is needed to put up such a structure, and one of this, or

similar form and make, might well be erected in a multitude of gardens. The more rustic the posts and other work, the better. (1859)

Straw Shelters, Stables, Ropes

The only abundant building material of the open prairie appears to be straw or hay. Shelters for all domestic animals are constructed of it. A few poles form a roof-support, and the straw is piled about and upon them. On the sides of the shed the straw is either simply a trodden down heap, trimmed with a hay-knife on the inside, or it is piled against rails. These are very warm sheds, see fig. 1, but they wet through, leak, and the straw rots and must be removed after a short time.

Fig. 1.

Accumulation of straw on the prairies is now prevented so far as the heading harvesters are used. These it is well known cut simply the heads of the wheat, leaving the straw as tall stubble. Still there are comparatively few headers in use, and straw heaps, new and old, dot the prairies far and near, or their burning illuminates the country by night.

Instead of the shelter now in vogue, better sheds might be built, using the same materials. Much of the tall stubble cut close to the ground is long enough to make most excellent and durable thatch, if well put on. A few bundles of wheat might be threshed out by hand, and the straw saved, or even the machine threshed straw might be used and answer tolerably well, if a sharp pitch be given to the roof. Thatching is understood by many immigrants and the principles upon which good work depends are so simple, that where beauty is not demanded, any handy man will make a tight roof after a little experience.

There are several methods of using straw to form the sides of walls of these stables. A convenient way is to set upright poles, about 8 inches apart, and draw wisps of straw around each, so that both ends of each wisp shall be outside. It is best to lay these in horizontal courses and beat down each course as it is laid, keeping it uniform and tight. As the filling in with straw progresses, there may be a split pole woven in once in three feet or so, to hold the uprights in place. The straw is finally to be raked down on the outside so as to shed rain well. This makes a tight, warm, and lasting wall. The inner side is quite even, and it may be sprinkled with mud if there is danger of the animals pulling out the straw to eat; see fig. 2.

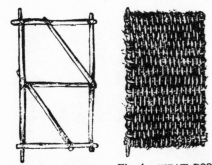

Fig. 2.—STRAW THATCHED SHED

Fig. 3.—DOOR FRAME.   Fig. 4.—STRAW DOOR.

It is a great convenience, where lumber is scarce, to be able to make expeditiously a good door or shutter of any kind. Constructed of straw a door may be strong, light, and tight. Tie, or wire together, a frame of round sticks—braced or stayed by cross-pieces to give requisite strength, as in fig. 3. This frame should fit loosely in the window or door-place, and one of the upright pieces should be strong enough to hang the door by. Then wind a straw-rope of 1½ to 2 inches in diameter around the longest way so as to cover the frame. Next, weave a tighter wound straw-rope, back and forth, plaiting the whole in a single mat, as in fig. 4. The strands on each side of the frame may be plaited separately, forming thus a double thickness

of the straw mat. We have seen affairs made in this way by the soldiers, and stuffed with straw as the weaving progressed, and when done they made very good beds.

Straw rope is made by twisting damp straw. Sprinkle a heap of straw the night before. All farmers should possess a set of center-bits and stock. Take a large center-bit and attach a stout wire hook to it and place it in the bit stock. Where the bit-stock is wanting, contrive some substitute. Two persons are required—one twists a loop of straw into the hook, fig. 5, and walks backward turning from left to right; the other remains at the straw heap and feeds fresh straw to the lenghtening rope. A sufficient length being attained, the rope is fastened upon a fence or between poles or trees until dry, when it will not untwist. (1864)

Fig. 5.—IMPLEMENT FOR TWISTING STRAW ROPE.

Making Straw Roofs

For twenty years I have been making roofs of temporary shelters with rye-straw. The straw should be cut when fully ripe, kept straight, and threshed clean by hand with a flail. All the grain must be removed from the straw, or rats and turkeys will make sad havoc with the roof.

I do not make the building more than 20 feet wide. Any kind of soft-wood poles will do for rafters, as small crooks will not show in the roof. Make the pitch very steep, at least half-pitch, or the ridge-pole ten feet higher than the plates of a 20-foot building. The rafters may be cut to project 1 foot, or if poles are used for rafters, small poles, say 2 inches in diameter, may be cut 2½ feet long, flattened at the upper end, and nailed to the rafter resting on the plate, thus making the desired projection. Nail a good solid small pole, say 1½ inch in diameter, on the end of the rafters, all around the building, splicing when necessary, then use inch poles for sheathing, nailing them on about 9 inches apart. The roof should be hipped at the ends, as it saves gables and looks better.

Commence the roof with a loose bunch of straw; compact it in the hands. Let the butts project about 3 inches over the outside, and tie it to the outside pole and to the one next above. This front row must be doubly secured, to prevent the wind from lifting it. Lay on another row of bunches, letting the butts come down even with the first, but

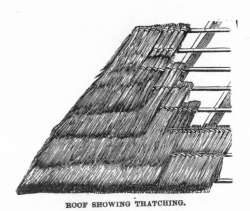

ROOF SHOWING THATCHING.

securing with but one tie to the third row of poles. Put on the third row of bunches, letting the butts come about 9 inches up on the two front rows, and proceed shingle fashion until the roof is completed. Finish the roof by binding a good supply of straw across the comb, running one or more wires along on each side, and sewing through with wire. For this one will need an assistant on the underside to return the wires through the straw. The corners, also, would be better secured with a little wire. Trim off the eaves with a knife. I have always used willow ties to fasten on the straw, but one can use twine or the kind of wire used by the binders.

An expert hand may make a good roof and tie it with straw. I have a shed made in this way 30 by 20 feet, used for storage, the whole cost of which was 50 cents for nails, and four days' work for two persons; the posts, poles, straw, and ties were home products. A roof thus made and not disturbed by stock, rats, or poultry, will last 12 or 15 years. But if used for stock, it must be put up high enough, so no animal can reach the eaves on a cold winter night. If for storage, and away from stock, the sides may be closed up the same as the roof, only not so thick. (1882)

The Greenhouse

A very good greenhouse on a small scale, may be put up at little expense. Success in the planning, and especially in the after care, depends mainly upon the taste, skill and constant attention of the owner.

The simplest form of a greenhouse is a well-lighted room in a dwelling, uniformly warmed in cool weather by a stove, or by a basement furnace, and kept constantly supplied with moisture by frequent sprinkling, and with fresh air by ventilators or windows. The main objections to such a room are: the difficulty of keeping uniform heat and moisture night and day, and that fact that if light be admitted by windows on only one or two sides, the plants will tend to grow one-sided, as they always develop most rapidly towards the strongest light. Frequent turnings of the different sides of the plants to the light will partially overcome the last named difficulty; while constant care and the use of steady burning fuel will measurably remedy the first named. The main thing at all seasons is to have only so much heat as is required to keep out frost.

Fig. 1.—OUTSIDE VIEW OF A BAY WINDOW.

A bay window, as it is commonly called (figs. 1 and 2) makes a good greenhouse on a small scale, if it is connected with a warm room, and is not on the north side of the house. It adds very greatly to the value of such a window to cut it off from the inside room by glass doors, as shown in fig. 2.

Fig. 2.—INSIDE VIEW OF A BAY WINDOW.

These enable one to readily admit warm air from the room, or to shut the plants off from an overheated atmosphere, and at the same time to admit external air, without introducing an unpleasant current into the house.

Immediately connected with this part of our subject are the various glass cases for the table, or window bench. Fig. 3 is a

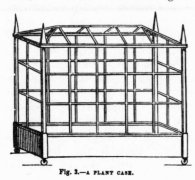

Fig. 3.—A PLANT CASE.

wooden box, 5 to 6 inches deep, lined with zinc, with a hole and plug, to draw off surplus water. The corner frames are of wood, about two inches square, or proportioned to the dimensions of the case, which may be of any size desired, from one foot up to half a dozen feet in length, and of proportionate width and height. The sides and top are covered with window glass set in sashes. A side door and an opening at the top to admit of ventilation when needed, complete the structure, except the painting inside and out, which may be of any color; green is usually preferable. Any carpenter can get up such a case, and it answers as a very good substitute for a diminutive greenhouse. The rules first given above, in regard to the light, moisture, warmth and ventilation, of course apply to all these structures for plants. (1881)

A Lean-To Greenhouse

A greenhouse may be placed by itself, or it may adjoin a dwelling or other building. The simplest form is a lean-to, on any side of a building except the north. Fig. 5 gives an outline form. The general architecture and form should harmonize somewhat with the building against which it is placed. To economize heat and labor, it should stand nearly on the ground level, if a dry soil underneath can be secured.

The foundation, carried say 20 inches above the ground, may be of stone or brick work; or strong posts of durable wood may be set in the ground and covered on outside and inside with boards, filling between with any dry substance, such as coal dust, tanbark, shavings, sand, or dry loam. The length of the structure may be anywhere from 15 to 500 feet. The highest part of

Fig. 5—LEAN-TO GREEN-HOUSE.

the roof should be about 10 feet inside measure, above the ground floor; and the lower side 4½ to 5 feet. If the foundation wall be 20 inches high, the sashes on this side will be 2½ to 3 feet high, after allowing for sill and plate.

The cheapest plan for warming a lean-to greenhouse is to have it connected by a door directly with a living room of the dwelling which can be kept warmed by a stove during cold nights as well as in the day time. A second, and still better mode of heating economically, may be adopted when the dwelling is warmed by a furnace. The hot air pipe from the furnace should enter the greenhouse at the bottom, and if convenient, in the part most exposed to cold. A broad pan of water set just over the opening of the pipe will serve both to spread the rising warm air, and to supply moisture. (1881)

Spring Houses

The points necessary to look at most particularly in constructing a spring house are the coolness of the water, the purity of the air, the preservation of an even temperature during all seasons, and perfect drainage. The first is secured by locating the house near the spring, or by conducting the water through pipes, placed at least four feet under ground. The spring should dug out and cleaned, and the sides evenly built up with rough stonework. The top should be arched over, or shaded from the sun. A spout from the spring should carry the water into the house. If the spring is sufficiently high, it would be most convenient to have the water trough in the house elevated upon a bench, as shown in fig. 1. There is then no

necessity for stooping, to place the pans in the water, or to take them out. Where the spring is too low for this, the trough may be made on a level with the floor, as in fig. 2.

The purity of the air is to be secured by removing all stagnant water or filth from around the spring, all decaying roots and muck that may have collected, should be removed, and the ground around the house be either paved roughly with stone or sodded. The openings which admit and discharge the water, should be large enough to allow a free current of air to pass in or out. These openings should be covered with wire-gauze to prevent insects or vermin from entering the house. The house should be smoothly plastered, and frequently whitewashed with lime, and a large ven-

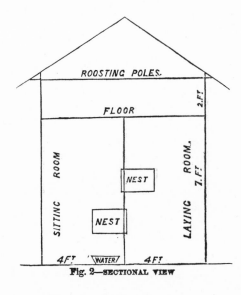

tilator should be made in the ceiling. There should be no wood used in the walls or floors, or water channels.

An even temperature can best be secured by building of stone or brick, with walls 12 inches thick, double windows and a ceiled roof. In such a house there will be no danger of freezing in the winter time. The drainage will be secured by choosing the site, so that there is ample fall for the waste water. The character of the whole building is shown in fig. 3. (1874)

The building, fig. 1, is 8 by 12 feet on the ground — the side walls 9 feet high to the eaves. A partition runs lengthwise through the middle, from the ground up to a floor which covers all the interior, at the height of 7 feet. Two doors in the gable end open, respectively, into the two rooms thus formed. There are two tiers of nests, containing ten in each, all of which are movable boxes or drawers, so placed as to slide freely through the partition, from one room into the other. The right hand, or "laying room," has an ordinary opening through which the hens have

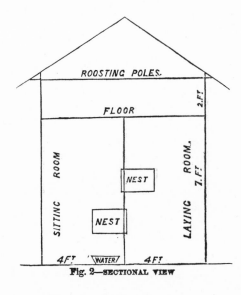

Fig. 2—SECTIONAL VIEW

ready ingress and egress, but the "sitting room" is closed to all but the "sitting members" and their human visitors. In this room, supplies of food and water are constantly kept, accessible to the sitting hens. Above the floor, or in the second story, are the roosting poles, to which access is had from the outside, as represented in fig. 1. The sectional outline, fig. 2, will illustrate the interior construction. (1863)

Such a root-house is shown in the engraving. An excavation about three feet deep is first made, and the log walls built around it. The walls should be double, at least a foot apart, between which earth is placed; the roof is made of timbers, with earth above, thick enough to keep out the frost. (1882)

Adobe and Concrete Buildings

The name *adobe* is applied to building material in the form of bricks, but unburned. It is not necessarily clay, but it must be of a clayey or loamy nature, and so firm when dried in the sun as to be easily handled, and to sustain considerable weight without crushing. This is distinguished from concrete by its containing no lime or cement, and being always used in the form of bricks.

The concrete is a mortar of sand and lime, usually hydraulic lime or cement, in part at least, especially such as is used for foundations, in which the proportion of cement preponderates. This is laid up in the walls as it is to remain, and the largest possible quantity of gravel and larger stones is worked in. The present high price of building materials leads farmers to look anxiously for some good substitute for stone, brick and wood. The need for the dissemination of knowledge on this subject is great, especially throughout the prairie states. A correspondent in Iowa pleads for information, definite and full, about adobe building, which we shall try to give, and would be glad to give also particulars of the experience of any of our readers who have tried, and either approve or condemn this material. (1867)

A Cheap Poultry House

A correspondent "Lex.," contributes the accompanying description of his poultry house.

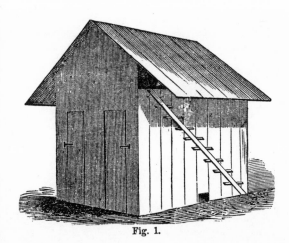

Fig. 1.

A Log Root House

It is not well to delay the building of a root-house until the time for harvesting the roots. The essential points in a good house for the storing of roots are: protection from the frost, dryness, and proper ventilation. For those who live in a wooded locality, a house built of logs is cheap and very satisfactory.

A LOG ROOT-HOUSE.

VII Fencing & Bridges

Cedar-Brush Fence

This is quite a common fence in Virginia, and is occasionally seen in New Jersey and Delaware. If well built, it is a good and durable fence. It is most usually made in this way:

First, throw up a ridge of earth about a foot above the level, and in this drive stakes on a line two to three feet apart, and then wattle in the cedar limbs, beating them down with a maul as compactly as possible. This fence will last good as long as the stakes endure. Some leave the stakes about a foot above the fence at first, and drive them down as they decay, adding more brush at the same time, and thus the fence will last fifteen or twenty years, with less repairs than a common rail fence. (1850)

Buckthorn Hedges

This beautiful shrub was first used in this country for growing live hedges by Mr. E. Hersy Derby, of Salem, Massachusetts. He first commenced propagating it by cuttings in 1806, from an individual tree which stood in the venerable Dr. Holyoke's garden in that time.

Since that period up to the present time, he has constantly multiplied the species by cuttings and seeds, and independently of ornamenting his own grounds, he has sent plans and scions to nearly every state in the Union. He has at present for sale at his nursery, several thousand vigorous plants at $3 per hundred, with a reasonable discount where large quantities are purchased.

This shrub flourishes best in rather a moist soil, although it will thrive in any soil that is adapted to the culture of garden vegetables. It is exceedingly valuable for live hedges, on account of its being able to resist any degree of climate in the United States, and of sure growth whenever transplanted. It remains green many weeks later than the English hawthorn, and in the vicinity of Boston, retains its foliage long after the fall of snow. In consequence of its medicinal properties, it is not attacked by any insects, nor devoured by any of our domestic animals.

A hedge may be set in front of a dwelling, or used as an enclosure for ornamental grounds. As the plants will grow to the height of 12 or 15 feet, they may be trained over an arch or trellis, and form a beautiful, densely-shaded arbor or walk.

It is preferable to set out the hedges in trenches from 3 to 6 inches deep and 18 inches broad. The plants may be arranged in two rows 10 or 12 inches asunder, and set 6 or 8 inches apart, placing those of the second row opposite the center space of the first. (1844)

American Holly Hedges

The American holly is the sturdiest and best armed tree in the world, flourishes in all locations, and presents, in a few years, a barrier which defies the inroads of man and beast; and in its red berries furnishes such a grateful repast to the birds, during the winter and spring months, that they, too, are enticed from depredations on the spring crops, and seek shelter and bounty in its thick impenetrable covert.

A great error has always prevailed in regard to the uncertainty of transplanting the holly, but from our experience and observation, there is no tree more easily and successfully removed. We have seen them succeed when trees four inches in diameter were taken up, but they had been entirely divested of their tops. Our friend, Mr. J. C. Singleton, residing near Columbia, S.C., has a garden hedge in front of his house, of several hundred small holly trees, transplanted from the adjacent woods, and lost not a single plant where they were entirely undisturbed.

98

The holly is easily propagated from the seed, which must be subjected to the following process, which makes them vegetate freely: In the fall, after frost, take a large quantity and bury them in a heap, in a soil not too moist. 'Let them remain until spring, when upon their being planted in drills, in finely-prepared soil, they come up quickly, and a number of them make plants enough fro transplanting the first year. It is, however, best to leave them till they are two years old, when upon planting, each plant should be cut off to within two inches of the ground. They should be planted in double rows, eighteen inches apart — the trees being broken in the ranks, and distant from each other in the rows, about fifteen inches. After the holly is well set, it requires no artificial culture. It is best, however, to shorten the plants down every year until the whole wall is a stout barrier of living trunks, and then it may be left to the care of nature.

A holly hedge, or indeed any other, should be planted in the soil prepared with a view to support the growth of the plants for years to come. The best plan is to dig a wide, but shallow ditch, into which, after throwing the top soil, place vegetable mould, muck, animal manure, and, in fact, all such materials and rubbish, usually found about the farm, which is conducive to the growth of trees. When the trees have taken root, and in order to make a good, sound, enduring wood, they should receive a dressing of wood ashes, or old lime. With these precautions, every farmer could have a good hedge in ten years, for the same labor and cost that it would take to keep a rail fence on the land for that time. (1848)

The Osage Orange for Hedges

The Osage orange (*Maclura aurantiaca*), known also by the names of Osage apple, bow wood, and *bois d'arc*, is indigenous to Arkansas, Texas, and Upper Missouri, and may be safely cultivated for hedges or ornament wherever the Isabella grape vine will thrive, and mature its fruit in open air.

In its natural habitat, the Osage orange forms a beautiful, deciduous-leaved tree, often growing to a height of 25 to 30 feet, with a trunk from 12 to 18 inches in diameter; and in very favorable situations, it sometimes attains double these dimensions. The general appearance of this tree greatly resembles that of the common orange; and when we view the beauty and splendor of its dark, shining foliage, large, golden fruit, and the numerous, sharp spines, which the branches present, we are strongly impressed by the comparison. The juice of the young wood, leaves, and fruit, consists of a milky fluid, of an acrid or insipid taste, which soon dries, on exposure to the air, and contains a considerable proportion of an elastic gum. The

fruit, however, in open culture, does not ripen its seeds north of Philadelphia.

The most important use to which the Osage orange can be applied is, for the formation of hedges; and there is no plant in our estimation, better adapted for this purpose, in any part of the country, where this tree will thrive. Apprehensions have been expressed by some, that, from its rapid growth, it will soon become too large for live hedges, which, it is thought will not endure for a great length of time. This, however, remains yet to be proved. We have no doubt, in our own minds, that if a judicious system be pursued, in trimming and heading down, they will serve an excellent purpose for twenty, and perhaps thirty years; for there are hedges of this plant in the vicinity of Cincinnati, which are ten years old, and have thus far proved perfectly hardy, very uniform, neat, and handsome in their appearance, and free from the attacks of insects or disease.

The Osage orange may readily be propagated by seeds, from which it will grow sufficiently large in three years to form a hedge. It succeeds best on land moderately rich, such, for instance, as will produce good Indian corn; but it will grow in almost any soil that is not too moist. The line of ground, intended for a hedge, should first be dug and well pulverized, say from 12 to 15 inches deep, and 2 feet wide, along the center of which the plants may be set at a distance of one foot apart.

The seeds, before sowing, should be soaked in water, in a warm room, for four or five days; or they may be mixed with equal parts, by measure, of sand, and exposed a few weeks, in open boxes, to wintry weather, on the sunny side of a building, in order to freeze and thaw. It is preferable to sow them early in the spring, in a garden or nursery, where they will shortly germinate and form young plants. These should carefully be weeded or hoed during the first season's growth, and transplanted in the hedge line in the month of March or April the following year. (1848)

How to Make a Pole Fence

"W.L.T.," Mount Hope, Wis., gives the following method of making a pole fence, which may be usefully adopted where the timber is too small to be split. The method may also be applied in part to the preparation of split rails for a post and rail fence of the ordinary kind.

The poles are cut 10 feet long, the posts being set 9 feet apart. Each end of the pole is hewed flat, so that it can be nailed to the post. For convenience in hewing them, the following contrivance is used. A pair of blocks are procured, and made in a "bed," by nailing strips to them crosswise. These

blocks may be kept exactly so far apart that they will serve as guides for trimming the poles to their proper length. Notches may be cut into each block, in which the pole to be trimmed is placed.

A "horse" is used to hold the pole firmly, while it is being trimmed. This is made of a heavy pole, 20 feet long, and a foot thick at the butt end. At the thick end two legs, 3 feet long, and at the other end two pins, 8 inches long, are inserted. The pins are placed so that when they are made to rest upon the pole, they grasp it and hold it firmly. While the pole is thus held, the ends are without difficulty trimmed and hewed, as may be desired.

The ends of the poles are nailed to the fence posts, and to keep them level with one another, each end of the pole to a different side of the post, or each panel of fence may be nailed to opposite sides alternately. (1875)

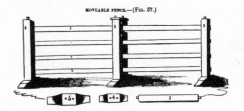

MOVEABLE FENCE.—(FIG. 57.)

Moveable Fence

When I visited the beautiful farm of J. F. Sheafe, Esq., at High Cliff, Dutchess County, N.Y., the past summer, among other things which struck my attention there, was a moveable fence; and, as it is both handsome and cheaper than the usual method of making these, as in England, of hurdles, I have thought a description of it might not be without interest to those desiring to put up something of this kind.

1,1,1, are posts made of joist, 3 inches by 4, and 4 feet long, with 2-inch pivots in each end, to fit into the holes of the cap 4 and step 5; 2,2,2,2, are slats 12 feet long, 4 inches wide, and 1-inch thick, nailed to the posts 1,1,1, making each four slats with their posts, an independent length of fence; 3, is a post the same as 1,1,1,; 4, is a cap of white oak plank, 12 inches long, 7 inches wide, and 2 inches thick, with 2-inch auger holes bored through it at proper distances, into which to insert the top pivots of two of the posts when set upright, to form the fence; 5, is a step, a piece of plank the same as the cap 4, but 3 feet long. It has two holes for the lower pivots of the posts to fit in, corresponding with those in the caps.

Two boys can take down or put up 40 rods of such a fence in an hour; and it is so light, that it can be transported to any part of the farm with great ease. When neatly made and painted, it has a handsome appearance, and proves an effectual guard for all orderly stock. By taking it down in the winter, and putting under cover, it will last a long time. It may be made very cheap by using slabs for the caps and steps, and split poles for the slats.

Mr. Sheafe showed me another kind of moveable fence with pickets, but I rather prefer this as being more simple. (1843)

Fig. 1.—AN OLD FENCE ROW.

Fig. 2.—A GOOD FENCE.

Board Fences

Posts. The most durable kinds of wood that we have are locust, cedar and oak. The variety called the swamp white oak is considered the best, though the upland white oak is often used. Of cedar, the red is far preferable to the white. The locust is better than all others; and here I would observe that, if farmers would pay more attention to the raising of the locust, they would soon find it much to their interest. The locust grows rapidly; and if set out by the road side, or on rough, stony, or otherwise unprofitable parts of the farm, they would in a few years be amply repaid for the labor and expense. On the road side they are ornamental, and fragrant in the season of blossoms, and give a pleasant shade. Fifteen years' growth will make from three to six or eight posts to each tree—the most durable of any timber in our country. As evidence of its durability, I may mention that a friend of mine visited his native town in Massachusetts, and there examined a locust post that had been in ground, according to tradition and actual knowledge, for 70 years, and it was still in pretty sound condition.

For a good strong fence, and none other should ever be made, the posts should be set not to exceed 6 feet apart. If the boards are only 14 feet long, then one post in the center

will do; but if they are 16 or 18 feet in length, two intermediate posts should be used. Two and a half feet is a proper depth to set the posts, and after they are thus set, and before the boards are put on, an embankment should be thrown up around the posts of from 6 to 10 inches high at least; more than this would be better. The surface at the center of the embankment, or along in a line with the posts, should be made level and smooth; this can be done before putting on the boards, much easier, quicker, and cheaper than at any other time. The team can pass between them, and draw the plow without hindrance. The benefit of this ridging is obvious; it secures the posts to a greater depth in the soil; it makes an even surface for the bottom board; it stops up all hog holes, and causes the water to pass off more freely from the fence in the ditches on each side. The ditches should be about two and a half, or three feet from the center, each way. Having set the posts and leveled the ridge of embankment, then proceed to face the posts (if split from the log) with a common narrow ax. As they are held fast in the ground at one end, a man, with a little mechanical ingenuity and a good eye, can easily and quickly do this with sufficient accuracy. If the posts be sawed, no other fencing will be necessary.

In nailing the boards upon the posts, the most common practice is to break joints, and, if no gaps are designed to be put on the fence, this, undoubtedly, is the better plan; but to secure a durable and strong fence, it is better to make each panel separately; select the broadest, widest, and best posts for the ends of the panels, and set and secure them in the ground first. By doing this you will be more likely to get the fence straight, and less care will be needed in placing the intermediate posts. An inferior intermediate post will answer the puspose, perhaps, as well as the best. If your ground is uneven, as up hill and down, a fence, when completed, will have more symmetry and beauty.

Height. Four feet above the embankment is high enough for any fence. This height is fully equal, in effect, to one of four and half or five feet, when built in the ordinary manner.

Width of boards. There is, I conceive, a stereotyped error in graduating the width of the boards; that is, putting a wide one for the bottom, and narrower ones as you approach the top. The better plan is to have them all of the same width, and that width five or six inches only; it is desirable to have frequent open spaces, that the wind may pass through more readily, and not to have so great a width of board resting upon the post at one point, holding moisture and causing it to rot much sooner. For this reason, also, split posts, as a general rule, are preferable to sawed ones. Select one having a pretty wide face, for the ends of the panels to be nailed to, and all the others trim in such a manner as to

have as little surface touching the board as possible. One inch surface, or even less, to nail to, is better than four or five. Battens also induce rot; if used at all, they should be put only over the ends of the boards where the panels come together, and be only about four inches wide. For an extra strong fence, a board should be nailed on each side of the posts at the top, and after all the posts are sawed off level, put a cap on, and nail it securely to both boards, as well as to the top of the poses, observing to break joints by placing the center of the cap over the post where the panels meet; this gives the whole strength of the cap edgewise in keeping the fence stiff at the top. In doing this, it will be perceived that the first cap must be sawed in two in the middle, and put on first, then all the others at full length will break joints over the proper post. It is not necessary that the ends of the caps should come together over the end of a post; it would be best not to have them do so, for the reason that when a whole board is made to cover the end of the post no water can get to it, to cause it to decay. Where the ends of the caps come together, more or less water will in time pass through the joint. One word as to putting on the caps. Before commencing the operation, see that the top of the fence is straight; if any variation from a straight line is noticeable, crowd the post in or out as the case may be, and secure it there by temporary braces. After this is done, nail on the first cap of half length, then put on the second cap, and let the end of it lap over the first, half an inch or more, and nail it securely also, except three or four feet from where the cap is. Then pass a saw through both caps where they come together, giving it a slight inclination from the perpendicular or right angle, and pass it through both boards, and when the short pieces are removed, the ends come together with a perfect joint; then nail them to the very ends.

The strength and practical benefit of a fence made in the way just described has been fully demonstrated within my own knowledge and I can recommend it in the confident belief that it will fully answer the most sanguine expectations.

If the ground be wet and springy, or a heavy clay soil, in which the posts when set will be liable to heave, and eventually be thrown out of the ground, the better way will be to dig a ditch at once, two feet and a half deep, and set the posts into it, at proper distances, and fill up the ditch with small stones; or, if they are not convenient, lay in the tile pipes in the same manner as for a common tile underdrain. Or, if neither can be had, brush or bits of rail, or anything placed in the bottom of the ditch that will lead off the water, will answer the purpose; observing, at the same time, to have a free outlet for the water at all points where it may collect in the lowest ground. This, with the small embankment, and its ditch on each side, will effectually secure the posts from ever being raised out of ground by the action of the frost. The ditch can be made almost, or quite as cheap, as to dig the post holes separately, and at the same time you will have a good underdrain.

With reference to the kind of lumber most suitable for fence boards; pine is the best, hemlock next, and these should be sawed full an inch thick. But when oak, beech, hickory, maple, ash, or elm is used, the boards should be sawed much thinner. Five-eights or three-fourths of an inch is thick enough for the reason that all these different kinds of wood will warp very much in the sun; and if thick, say an inch or more, there will be so much strength in them that they will draw the nails or split in warping. But if thin and narrow, they will keep their places much better, and are sufficiently strong for all practical purposes. (1860)

Fig. 2.—THE POSTS.

Fig. 3.—THE FENCE COMPLETE.

Fig. 2.—FRAME FOR MAKING FENCE.

when nailing them to the rails. These panels can be made in the shop or on the barn floor at odd times, and piled away for future use. For pigs or poultry, nail a wide bottom board around on the inside of the enclosure after the fence is in position. (1883)

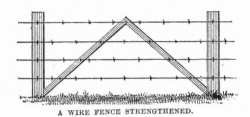

A WIRE FENCE STRENGTHENED.

A Simple Portable Fence

There are many places on the farm, and especially about the barns, where a few lengths of portable fence are a great convenience. It may be that an inclosure is needed for some sheep, or a pen is required for a calf, etc. In such cases a hurdle fence of some kind, one that may be cheaply and easily made, and that, when not in use, can be packed away in small compass, is an important part of the outfit of a farm.

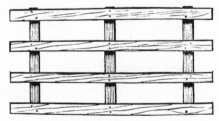

Fig. 1.—FORM OF SINGLE HURDLE.

The accompanying engravings show a simple fence of this kind. Fig. 1 gives the form of a single hurdle, or length, made of four strips of inch board, 12 feet long, and of a width depending upon the purpose for which it is to be employed; a fence for small stock will need wider boards than for turning large animals. The perpendicular pieces are three in number, the outer ones set far enough back from the ends to allow the supporting posts, fig. 2, to take the position shown in fig. 3. The next to the upper strip rests in the crotch of the posts, while the bottom strip fits into a notch in the cross piece, near the feet of the posts, thus making the fence firm and secure. The hurdles and supports, with the manner of setting up the fence, can be understood from the engravings. (1880)

A Cheap Fence

To those who desire an inexpensive fence for confining poultry, or for protecting their garden and grounds, I would recommend the one described below, which I have used for the purposes above-mentioned, and also for a movable fence about the farm. It is really an ornamental fence compared with most of the fences we see around farm houses.

Fig. 1.—PANEL OF PICKET FENCE.

The panels, fig. 1, are 16 feet long and are composed of 2 pieces or ordinary 6-inch fencing, for top and bottom rails, with lath nailed across 2½ inches apart; the top ends of the lath extending 10 inches above the upper edge of the top rail.

Posts, 3 or 4 inches through at the top end, are large enough and after sharpening well, can be driven into the ground, by first thrusting a crow bar down and wrenching it back and forth. A post is necessary at the middle of each panel. Both rails of the panel should be well nailed to the posts. These panels may be neatly and rapidly made in a frame, constructed for that purpose.

This frame, shown in fig. 2, consists simply of 3 cross pieces of 6 by 6, 4 feet long, upon which are spiked two planks, 1 foot wide, and 3 feet apart, from outside to outside. Four inches from the inner edge of each plank is nailed a straight strip of inch stuff, to keep the rails of the panel in place while the lath are being nailed on. Against the projecting ends of the cross pieces, spike 2 by 6 posts 12 inches long; on the inside of these posts nail a piece of 6-inch fencing, to serve as a stop, for the top ends of the lath to touch

Fencing & Fences

This important subject comes home to every owner of a farm or of a village lot that requires enclosure. At a meeting of New Hampshire farmers, several gentlemen publicly offered to sell their farms for less than what the existing fences on them had cost.

During 38 years past the *American Agriculturist* has had not a little to say as to the uselessness of a good deal of fencing. But much will be needed wherever live animals are kept, and we propose now to direct some effort to reducing the cost of fencing generally. If this one item in the United States can be reduced only one-fourth, the saving will amount to $500,000,000. In the older States, there are an average about two miles of fencing for each 100-acre farm, costing about $1 a rod, or $640.

Many of our readers have the barbed fences. We earnestly invite accounts of their experience and observation, *pro* and *con*. We have some letters denouncing barbed wire fences as too cruel to be permitted, and insisting that they should be prohibited by law along highways and railways. Others speak of their introduction as of inestimable value.

There is a pressing need of new and improved forms of iron fence posts. The simplest form now is an iron bar, or an iron tube, with holes or staples for attaching the horizontal wires. The same weight of iron is stronger in the form of a tube than in a solid bar. But cast iron will be likely to be used generally. After securing sufficient strength of post above ground, the next points to be aimed at are light weight, on account of expense, and breadth of resisting surface in the soil to prevent swaying in any direction. Nonliability to heaving by frost is also an important point.

Many attempts have been made to imitate the natural thorn-armed plants, by affixing barbs to wire, and over 30 different patterns have been devised. We have some 25 specimens, the exact size and form of which are shown in the engravings below. Nearly all the fence wires here described, are now made of steel, and most are supplied either painted (or japanned) or galvanized (coated with zinc).

The engravings are so exact that our readers can almost decide and make their own selection. We confess to being yet in the condition of the learner—but an earnest, investigating one, in behalf of the public. We started with a prejudice against any of the severer forms of barbs; yet some manufacturers say the demand is for the longest barbs, especially at the far West. We intend to learn more and say more on this, especially when we hear further from our readers experienced in their use.

No. 1 is probably the first attempt at "barbing;" said to have been made by an Iowa blacksmith some 20 years ago. It is a common horseshoe nail bent around the main wire, and held in position by binding it with a small wire as shown.

No. 2, the "Cleveland Barbed Wire." The points stand out at nearly right angles, which could not well be shown in the engraving perspective. The same barb can be applied to double or triple wires. The names of the manufacturers unknown to us.

No. 3. Engraved from a specimen; distinctive name unknown. These barbs also stand out at right angles, that is, in four directions. We are not aware that this form is now being manufactured.

No. 4, the "Kelly Steel Barb Wire," with two-point barb stamped or cut from sheet metal, and pierced to string upon one main wire and be held in place by the other. This variety is manufactured by the "Thorn Wire Hedge Co.," we believe.

No. 5, "Roberts Barbed Wire," engraved from specimen; if now manufactured, the makers are as yet unknown to us. The two-pointed barb is a cast double-pointed piece of metal, with a deep groove around its middle, where it is grasped and held somewhat firmly by the twisted main wires.

No. 6, the "Crandall's Barbed Wire," is simple, and this barb may probably be put upon a single wire, like No. 2, but when there are only two barbs, they need be set closer together than for the four barbs. Made by the Chicago Galvanizing Co.

No. 7, The "Sterling Barbed Wire," has a single pointed wire bent firmly around one of the main wires, and locked over itself; it presents the two barbs, pointing in the two opposite directions. This wire is manufactured by the North-western Barb Wire Co.

No. 8, the "Bronson Barbed Wire." The two barbs are formed by cutting one of the running wires, and bending and locking the ends. We do not know the manufacturers.

No. 9. The "Glidden Steel Barbed Fencing." The barb has a close double turn around one of the main wires. To reduce the objections to dangerous points, this variety has for some time past been made with the barbs materially shortened. Manufactured by the Washburn & Moen Mfg. Co., as per advertisement elsewhere.

No. 10, the "Three-pointed Stone City Steel Barbed Wire," has a 3-pointed piece locked between the two main wires. Made by the Stone City Barb Wire Fence Co.

No. 11. "The Steel Barbed Cable Fence," (Fentress & Scutt's patents) is similar to No. 10, but with 4 barbs on one solid piece. Made by the Illinois Fence Co. Two other barbs, somewhat similar to these, are made by H. B. Scutt & Co.

No. 12. "Spiral Twist, 4-pointed, Steel-barbed Cable Fence Wire" (Watkin's patent). The barbs are on one solid piece of metal, which is bent to conform to the twist of the main wires, and is thus held fast. Made by Watkins & Ashley.

No. 13, the "Quadrated Barbed Fence," is explained by the engraving. It has well-fastened barbs certainly. Made, we believe, by Pittsburgh Hinge Co.

No. 14. The "Iowa 4-pointed Barbed Steel Wire," (Burnell's patent). The form and structure are plainly shown in the engraving. The wire of the barbs passes twice around and between the main wires, but so loosely as to yield a little. It is made by the Iowa Barb Steel Wire Co., both in Iowa and New York.

No. 15. We give the name "Lyman Manufacturing Co. Barbed Fence" to this, from the company reported manufacturing it.

No. 16. The "Allis Patent Barb," is all the name we have heard for this. Our specimen is a solid piece of the form shown in the engraving herewith.

No. 17, the "American Barb Fence." A central, wire closely sheathed with a continuous metal strip, with its edges cut in the form of barbs, turned out in all directions, the points one inch apart. The whole is covered and saturated with paint or zinc, firmly cementing the outer and inner metal. Made by the American Barb Fence Co.

No. 20 is a partial section of No. 19, which is introduced to show how it is twisted in putting up.

No. 21 is from the actual piece we brought from a fence at Beloit, Wis., (sketched from memory last month, not quite correctly). Objectionable from its too sharp barbs, and the expensive waste of cutting out the metal between the barbs.

No. 22. "Scutt's Patent Tablet Wire," provides a row of wooden blocks on the middle or top, large enough to be readily seen by animals approaching.

No. 23. "Scutt's Lock Center Barb," a modification of No. 11, as is shown.

Gates with Wooden Hinges

A subscriber in Tasmania sends us sketches and following description of a handy gate with wooden hinges: "Bore a large auger hole through the gatepost, both above and below; make the sockets of tough wood dressed to drive through the auger hole, and put in a pin on the other side. The sockets for the gate are, of course, bored, and the ends of the gate head rounded to fit, with a shoulder. This plan is superior to a socket placed in the ground which usually holds the water and soon rots." The arrangement of the parts described is shown in fig. 1; the socket seen from above is given in fig. 2. The same correspondent says: "I have constructed a gate, full sized gateway, say 10 feet, intended to be merely temporary, but afterwards properly hung, and still as good as ever, four

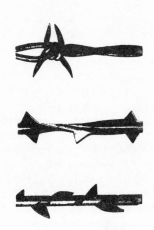

No. 18, the "Brinkerhoff Steel Strap and Barb," of the form and size shown in the engraving. Manufactured by the Washburn & Moen Mfg. Co.

No. 19, we call the "Brinkerhoff Improved." This is the Brinkerhoff strap, same as No. 18, but the barbs, which were prepared in this form at our suggestion and request, project only about one-fourth of an inch, and the edges are so inclined that they will not catch and tear the skin of animals unless pushed horizontally under heavy pressure, nor will they injure clothing swinging loosely against them. Others claim strongly that the barbs are not long enough or sharp enough to repel cattle. That is one of the points to be decided. Indeed, the leading question in the discussion now is, how much, or how little barbing is best, taking into account effectiveness, and safety for animals.

Nos. 24, 25, 26. "Lord's Rotary Barbs." Shown are three different forms of these, in which the barbs are arranged to turn or revolve loosely in the wire. The claim is that they yield to the motion of the animal, and lessen the tearing. (1880)

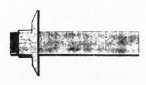

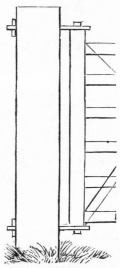

Fig. 1.—GATE POST WITH WOODEN HINGES.

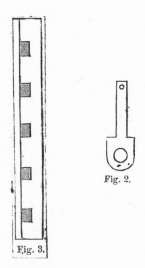

Fig. 2.

Fig. 3.

years' old and continually in use. It was made of pieces of an old roof, rafters and battens, and did not take more than half an hour to put it together. The rafters, for heads, here checked out the width and thickness of the battens with a saw, the checkers knocked out with a chisel, the battens inserted in their proper places, and a batten nailed over the head. The nails were clenched, and the whole gate simply braced. The size of rafters was 2 by 3, and the battens 1 by 3 inches." A single head is shown in fig. 3. Our subscribers in Tasmania and Australia are quite numerous and we hope others will give our readers the benefit of their experience. (1882)

Fig. 2.—A GATE WITHOUT HINGES.

A Gate Without Hinges

A subscriber in Pennsylvania, sends a description of a gate, which is without hinges. It turns upon a pin, of 1¼-inch round iron, fixed on a block of wood or a stone, under the heel-post, and a pin bent to a proper shape, fitted in the top of the heel-post, as shown at fig. 2. This method of hanging a gate is cheap, simple, and more secure, than to use hinges. (1876)

A GATE FOR FOOT-PATH.

Concrete Setting for Fence Posts

There is constant inquiry for some means of setting fence posts so that they will not heave by the frost. The following is suggested as offering at an expense of a few cents per post, an effective way. A hole is dug about as large as a flour barrel, but wider at the bottom than at the top, on two sides at least. The

post is set upon a stone laid in the bottom and the hole is filled up rapidly with concrete made of good hydraulic cement, mixed with half as much again sharp sand or gravel as would be used in making builders' mortar; and during the filling, as many clean stones, large and small, are thrown in as can be buried in the mortar. Posts thus set will be firm as rocks, and will not decay below ground. (1864)

Posts Lifted by Frost

The curious expansion of water in freezing, and of soils filled with water, has a telling effect upon fence posts standing in damp ground. The top soil around a fence post, if dry, or only slightly moist, does not affect the post during winter. But if the particles of soil are saturated with water, on freezing the whole expands an eighth, so that, when frozen eight inches deep, the post is lifted an inch out of the lower unfrozen soil. If the frost penetrates 16 inches, the post rises 2 inches. When the frost leaves, and the surface soil sinks back, the post remains 2 inches out of the ground. A few freezings of the surface will thus inevitably throw the post out so far as to render it useless, unless it is driven down every spring before the open space left at its bottom is filled by soil washed in. For this reason it is advisable to make the extreme lower ends of fence and other posts a little tapering, or at least to clip off the bottom corners so that they can be driven down more easily when lifted. We will say now in advance, that all fence and other posts should be examined early every spring, and those at all lifted be driven back with a beetle or

Fig. 1.—POST LIFTED BY FROST.

sledge hammer. A stitch in time will save nine, here. On naturally dry, or drained land, the above trouble will not be experienced, except at places where water flows through them from a higher to a lower level, thus keeping them wet, or very damp.

Stone fence walls, when standing on water-saturated soil and expanded uniformly on both sides, will not be affected. But almost invariably, especially if running in any direction but north and south, and often then, the soil under them will be more frozen or sooner thawed on one side than on the other, and thus they will be thrown out of perpendicular, and more or less disturbed. Only four inches of frozen wet ground under one side of a stone wall, and none under the other, will lift the frozen side half an inch, or enough to tilt five inches to one side the top of a wall five feet high and two feet thick at the bottom—enough to greatly disturb it, and ultimately throw it down. A wall set down two or three feet deep in the ground will be similarly affected, if water stands around its base. There are two remedies.

2. WALL ON A RIDGE. 3. WALL ON LEVEL GROUND.

One is to run a drain under or near the wall, deep enough to carry off all standing water about it, below the freezing point. The other is to raise the earth into a ridge before the wall is built, high enough to have it always dry. This is readily done by successive plowings, turning the furrows to the center of where the wall is to stand. The saving of foundation stone will far more than cover the cost of raising the ridge; and the fence will not need to be so high if standing on such a ridge, as animals will not jump it so well when they must spring from ground rising in front of them. (1882)

A Durable Stone Fence

The great objection to the old style stone fence, whether built single or double, was its want of durability. Unless the foundation was put below frost it was soon thrown out of line, and in a few years gaps were made in it every winter, and much labor was expended for repairs. It was, indeed, a better fence than one of rails, for the material never rotted, and it did not need resetting so frequently. It was always expensive, and would never have been so extensively built but for the convenient market it made for surface stones.

WALL OF HEAVY STONES.

The rock lifters, of which we have two, at least, mounted upon wheels, have introduced a new style of heavy wall that can be cheaply built, and will last forever. These machines will draw boulders deeply imbedded in the earth, weighing six or eight tons, and, with a single yoke of cattle or span of horses, will lay them in the bed of the wall. The smaller boulders are put in their position without any straining, or lifting, and a wall of five, or six feet high, embracing three tiers of stone, as shown in the illustration, can be laid by the team and two men. The interstices have to be filled up with smaller stones, and the large stones sometimes need blocking to make them bear perfectly. The largest boulders are five or six feet across, and this is the width of the wall at the bottom, as usually made. The stones next in size make the second tier, and the smaller ones form the caps.

A wall of this kind with a four-wheeled machine can be laid up for about $2.50 a rod, including the digging of the stones. The most expeditious method is to lay the stones as fast as they are dug, as this saves the labor of hitching on to them and raising them a second time. If properly laid, no frost will ever disturb such a wall, and it will last until the boulders crumble. Thus very rough pastures are economically cleared and fenced, and turned into smooth, productive meadows. The stone pulling is a very thorough subsoiling, and the effect is visible for many years. (1868)

Stone Walls

Quarry stones usually make better walls than loose boulders of any size. They are more shapeable, with flat surfaces, lie more compactly and evenly, and not so apt to fall, or be disturbed by frost, and where not too expensive, even if surface stones are at hand, are to be preferred. Yet surface and quarry stones do not always abound on the same premises, and the farmer is in most cases compelled to choose either one or the other.

The main rules which we shall lay down for a permanent wall are — 1st: A perfect drainage of the soil. If there be not a natural drainage, such as to permit no standing water on the surface; or at frost depth below, there must be an artificial one to such depth so as to allow the accumulating water to freely pass off under the wall, that it be not affected by its action in a frozen state. 2nd. If the soil be heavy or a stiff clay, holding water, an ample ditch dug down below frost level, should be thrown out, leading to lower levels and giving free passage for the water to a point where it can be turned off from the line of the wall.

Now, no matter whether the stones be surface boulders or quarried, this ditch should be filled nearly, or quite to the surface, with small or broken stones, compactly placed as a firm foundation. This preparation gives a perfect drain for the water beneath the small stones whether by rains, melting snows, or the issues of springs and keeps the adjoining soil in an equable condition winter and summer, while the earth at the bottom of the ditch is uninfluenced by winter's cold or summer, while the earth at the bottom of the ditch is uninfluenced by winter's cold or summer's heat — like the cellar walls of a house.

To make the best wall, the two sides are to be simultaneously built, with a line on each side to work by, and if two good wall builders can work together, one on each side, the same length of wall will be better and cheaper built than if but one work alone.

In first placing the lines let them be eight inches or a foot above the ground strongly held at each end by a peg driven into the ground, that the first tier of stones may be laid below them. The width of the foundation will depend somewhat on the kind of stone, and the height of wall when finished. Quarried stone being better shaped, requires less width at the foundation than boulders. In the former the lower strata should not be less than two feet for a substantial wall, four and-a-half feet high. If five feet high, a four inches wider base will be needed. The wall should not be less than one foot wide at the surface in any case, for a good farm fence. Without boulders, when stones are plenty, the foundation may be two and-a-half or three feet wide — and as much wider as you choose. The contraction of width should be uniform and gradual. Put in the largest stones first, making as close joints as possible, and if large gaps are left between, then fill them in evenly as possible with small ones, having an eye always to the importance of laying every stone so as to bind the wall together as firmly as can be done.

In every successive layer as the wall progresses upward, put in enough substantial binders; that is, stones reaching through from one side to the other, and let the last tier of binders alternate between the binders below so that there be a continuous bearing throughout the whole line. As the wall works upward the line is to be shifted, keeping it always a little above the tier on which you are at work, still making the batten or indrawing course uniform, and reducing the size of the stones gradually if possible, and holding on to the abundant distribution of binders but keeping the stones sizeable.

If the wall be on a side hill, a broader surface will be needed on the down hill side, but they must be laid flat, and worked in with a regular batten. If the line of wall runs down hill, the ditch or bottom should be made in short steps, so that the foundation be always on a level. A canting stone is easily displaced, therefore every one should lie in a horizontal position.

Each wall builder should have a short crow bar three and-a-half or four feet long, and a good stone hammer at his side for occasional use in knocking off an ugly corner, or splitting an ill-shaped stone to make a perfect job. As the wall approaches completion let the stones be more sizeable if possible, than further down, as they give it a more finished appearance; and if the top is to be a smooth and level surface, a tier of thin flat caps, if to be had, should make the finish. (1859)

CROSS-SECTION OF A STONE BRIDGE.

Stone Bridges

The most durable bridges are made of stone, which is preferable to wood for this purpose, even if the expense is one-third greater. A small, well-built stone bridge will last for at least 50 years, while one of wood will need renewing twice in that period, to say nothing of occasional repairs. When a stone bridge is out of repair, it is only necessary to adjust the displaced stone; with a wooden structure, the decayed portions are, for the purpose of repairing, worse than useless. For streams, where the water, during the heaviest freshets, will all pass through an opening 3 feet wide and 2 feet high, a stone bridge should be built. Larger bridges of this material are frequently met with, but they are expensive, especially if flat stones for covering are difficult to obtain. A section of a bridge, constructed wholly of flat stone, is shown in the engraving.

When it is difficult to obtain stones wide enough to cover the top, let a course of flat stone, along each side, project inward from 6 to 10 inches, as at *a,a*, and cover with the widest stone, as at *b*, laying on a course of smaller ones, to prevent earth from sifting through. A stone or small wooden bridge covered with 18 inches of earth, will outlast one having a covering of but 6 inches. Add to the depth of covering, even if necessary to elevate it above the roadbed at the sides. Lay a wall, or place large boulders at the upper end of the bridge, outside the channel, to prevent the embankment from washing away.

When the bridge is over a rapid stream or on a side hill, it is best to cover the bottom of the channel with stone. This should be done when building, and consists in placing flat stones at the lower end of the bridge, letting the stone wall rest upon them, or at least press firmly against them. One flat stone overlaps the one below, shingle fashion. Plank may be used instead of stone, being placed after the wall is laid, and the ends fastened by pins to a sill imbedded in the soil at each end. (1883)

Rustic Bridges

It often happens that a brook which traverses the farm or runs through the grounds has to be crossed by a path, and it affords the proprietor an opportunity to indtroduce an ornamental structure in the shape of a rustic bridge, which, if the location is well chosen, will add much to the attractions of the place.

To facilitate the crossing of small streams, we find on slovenly places a plank, or even a rail, made to serve as a bridge, but where the proprietor is more regardful of neatness and comfort there is usually a bridge of carpenter work. A bridge of rustic work is in much better taste than one carefully planed and painted, and can be made plain or quite elaborate according to the fancy of the builder.

The best material for this, as for other rustic work, is red cedar, as the wood is not only of pleasing color, and durable, but with a proper care in selecting, pieces may be found having a natural curve which adapts them to the use. In a bridge the work should be strong, and those parts in contact with moisture may be preserved by a coating of coal tar. The design may be graceful or express solidity, according to the size and situation of the structure. (1865)

A Wooden Bridge

Country bridges are always useful, but rarely ornamental. Designed for strength, appearance is sacrificed to utility. It is often the case, however, that the ornamental may be combined with the useful with advantage. In bridges of a certain character this is essen-

BRIDGE WITH WOODEN ARCHES.

tially the case. A simple timber laid across a stream as a foundation for a bridge, although the simplest and plainest form of structure, is far from being the strongest. The truss of lighter materials is stronger than a single heavy beam, while the arch may be made lighter yet than the truss, with a still further gain in strength.

We give a cut of a wooden arch, to be made of boards fastened together with nails and bolts, which may be built readily of materials always at hand, and needs no piece longer than twelve feet, even for an arch of 40 feet span or over. Nor is it necessary even to lay a center on which to build this arch. It may be built up on the ground, a foundation of stakes or posts being made on a level place on which to commence; or it may be built on a barn floor, if one of sufficient size, and when completed moved to its place and set up.

The mode of proceeding is as follows: We will suppose a bridge of 24 feet span is needed. The first necessity is to make the foundations for the arches. These should be built firmly of stone or timber, and well backed, and steps made to receive the the feet of the arches. If the bridge is to be 12 feet wide, three arches will be necessary. These are made of spruce boards, preferable as being elastic and tough, or wanting them, pine or hemlock will answer, and 12 inches wide and 1 inch thick. The form of the arch is laid out on say the barn floor, and a scantling tacked down for the base, with studs reaching it to the line of the arch. A board is then tacked to the end of the scantling, and bent round on to the ends of the studs, and tacked to them to hold it into its place; another board is put to the end of this, until the other end of the scantling is reached, and the figure of the arch is complete. Other boards are then placed over the first ones, and wrought nails driven through and clinched. The joints must in all cases be broken. Boards are nailed on in succession until a sufficient thickness is secured—12 to 20 inches, as may be needed for a bridge. (1876)

Fig. 1.—FRAME BRIDGE.

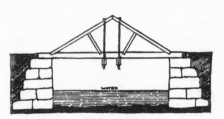

Fig. 1.—A SIMPLE FORM OF BRIDGE SPAN.

Fig. 2.—A STRONGER SPAN.

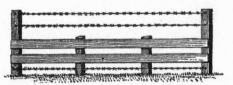

A WIRE GATE.

BOARD AND WIRE FENCE.

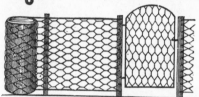

WINEGAR'S AUTOMATON GATE.

EWALD OVER,
MANUFACTURER OF
Automatic Farm Gates,
Indianapolis, Ind.

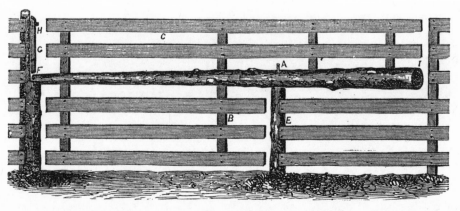

The Balance Gate.

VIII   Small Animals
& Horses

A PAIR OF BRONZE TURKEYS.—From the Flock of Wm. Clift, Mystic Bridge, Ct.—*Drawn and Engraved for the American Agriculturist.*

A Comfortable Bed for Animals

What man or beast does not enjoy it? Every wild animal, from the lordly lion to the insignificant mouse, bestows careful pains upon its nesting place. The universal instinct which prompts this ease, indicates that it is a matter of no small important in the physical economy. About one-third of an animal's whole life is passed in resting, and nature intended that during this time its condition should be most favorable for restoring and building up the organization.

In the care of domestic animals, kept for profit, this point is worthy of special attention. Comfortable bedding directly favors the increase of fat and muscle by helping to retain the animal heat, and also by adding to quiet and comfort. In this way a bundle of straw upon the outside may be equivalent to a feed of grain inside.

Horses are usually well cared for in this respect, with a view to keep their muscles in good order, as every tyro must know that sleeping upon a hard board will scarcely give pliancy to the limbs. But good bedding is of little less benefit to cattle. If it be doubted, experiment for two weeks with milch cows; give them comfortable litter the first week, and allow them to lie upon the frozen ground the second, then note the difference recorded in the milk pail; it will be very great.

Straw and refuse hay are generally used, and are well suited for bedding. Cutting into lengths of say six inches has some advantages, though it would hardly pay if required to be cut by hand. Where these can not be had cheaply, as is often the case in villages, an excellent substitute may be found in leaves. They possess one advantage over straw, in making the very best manure for gardening, when mixed with animal excrements. Spent tan bark, well dried, is another good substitute, also valuable as a mulch. A layer of dried muck, six inches thick, serves a good purpose for bedding. It is a most excellent absorbent, and will remain in good condition for some time without being changed. When well saturated it is just the article for the garden or the field. With proper care in furnishing abundant bedding for stock, a large accession to the manure heap will be made, sufficient of itself to pay for the trouble. (1862)

Fig. 1.—COVERED FEED BOX.

Feed Boxes

In figure 1, a box is shown firmly attached to two posts. It has a hinged cover, *p*, that folds over, and may be fastened down by inserting a wooden pin in the top of the post near *n*. The one given in fig. 2 may be placed under

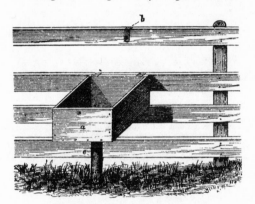

Fig. 2.—HINGED FEED BOX.

shelter, along the side of a building or fence. One side of the top is hinged to the fence or building, the bottom resting upon a stake, *e*. When not in use, the box may be folded up, the end of the strap, *b*, hooking over the pin, *a*, at the side of the box. A good portable box, to be placed upon the ground, is shown in fig. 3. It is simply a common box with a strip of

Fig .3.—PORTABLE FEED BOX.

Fig. 4.—FEED BOX INVERTED.

board, *h*, nailed on one side, and projecting about eight inches. When not in use, it is turned bottom up, as in fig. 4. The projecting strip prevents three sides of the box from set-

tling into the mud or snow. The strip is also a very good handle by which to carry it. Those who now use portable boxes will find the attaching of this strip a decided advantage. A very serviceable portable feed box is made from a section of half a hollow log, with ends nailed on as shown in fig. 5. By letting the ends project above the sides four or five inches, it may be turned over when not in use, and easily turned back by grasping the sides without the hand coming in contact with earth or snow. All feed boxes and racks should be placed under shelter during summer, or when not in use. (1882)

Fig. 5.—BOX FROM HOLLOW LOG.

The Horse of all Work

It is a difficult thing to determine exactly what should be considered a perfect horse — for the perfection of any domestic animal consists in its adaptation to the service required of it — be it ox, cow, sheep, swine, or horse. The English hunter is as near the type of a perfect horse of all work as can be found.

It is particularly necessary for farmers who breed horses to study their points, particularly with reference to sires. A breeding horse should be sound in all respects, except blemishes caused by accidents or violence, or by sickness which was neither inherited nor can be transmitted.

A good horse is moderately short-backed (12 to 34), and long below (17); round barreled and well ribbed up (16); rather high in the withers (12); having moderately sloping shoulders (12 to 23); a broad chest (24), a firm and muscular crest (11), a head well set on, lean and bony, with a clear,

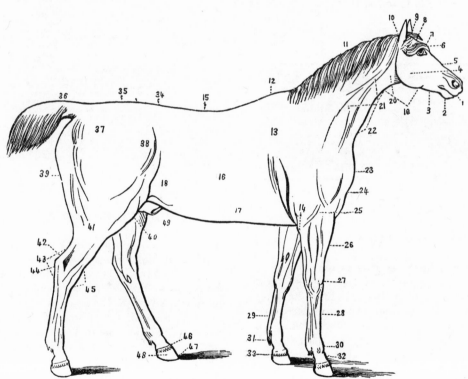

POINTS OF A HORSE.

NAMES OF POINTS WITH THEIR DISEASES.—1. The Muzzle including the chin, (2) lips and nostrils ; 3. the Jaw; 4. the Jowl ; 5. the Nose ; 6. the Eyebrow ; 7. the Eye, (the seat of various diseases causing blindness, weakness, flowing, etc., including glass eye, cataract, ophthalmia, exostosis, etc.) ; 8. the Forehead ; 9. the Ears, (affected by inflammation, causing deafness) ; 10. the Poll, (the seat of poll evil) ; 11. the Crest, (sometimes spongy and fat, and affected by mange) ; 12. the Withers, in which sometimes occurs fistula ; 13. the Shoulders ; 14. the Elbow, often injured, the seat of tumors ; 15. the Back (saddle galls, sit-fasts, warbles, etc.) ; 16. the Girth, (broken rib) , 17. the Belly ; 18. the Flank ; 19. the Throat (sore throat, laryngitis) ; 20. the Gullet ; 21. Jugular vein ; 22. Windpipe (bronchitis, injury from collar) ; 23. Point of Shoulder (contusions, lameness, and sprains) ; 24. Breast or bosom, (various internal diseases, chest founder, broken wind, etc., are located within the chest cavity) ; 25. the Elbow Joint ; 26. the Fore Arm ; 27. the Knees, (knee-galls, abscesses, broken knee—by white spots showing a tendency to stumble, knock-kneed, bow-kneed, weakness of the joint, etc.) ; 28. Cannon bone, (splents) ; 39. Tendons, often spongy, knotty, and the seat of wind galls and other injuries ; 30. Fetlock Joint, (the seat of swellings, osseous enlargements, stiffness, wind-galls, etc.) ; 31. the Fetlock ; 32. the Pastern (ring-bone, fracture, wind-galls, cracking of the skin, etc ;) 33. Heels, (grease) ; 34. Loins ; 35. Croup, (occasional dislocation) ; 36. Dock, (injury by the crooper, and mange) ; 37. Rump ; 38. the Hip, (bruises, wounds, and fracture) ; 39. Quarters ; 40. the Stifle, (subject to dislocation, bruises, sprains, and whirl-bone lameness) ; 41. Thigh or Gaskin, (sprain, string halt) ; 42. Hamstring; 43 and 45. the Hock, (blood spavin, bog spavin, bone spavin, capped hock, thorough-pin, curb, etc.) ; 44. the Hockbone ; 46. Corona ; 47 Toe ; 48. Walls of the hoof, (the Hoof. the seat of numerous diseases, must be explained on another occasion) ; 49. Sheath, (often foul).

bright, medium-sized, intelligent eye, an open, thin, broad nostril, clean muzzle, and small ears; his rump (35) should be straight, broad and full; his loins (34) broad; legs above the knee (27) or hock (43) long and muscular.

All the important muscles of the extremities are located above these points, and below them they should be short and bony, and the tendons (29) hard, and free from soft spots or excrescenses; the leg bones large, flat, and smooth; the pasterns (23) not too long or oblique; hoofs hard, clean, deep, (not flat) round (on the ground) and good sized. The knees (27) and hocks (43) should be broad and bony, the quarters (39) large, broad, and muscular, square when seen from the rear, the shanks from the hock to the pastern short, hard, and clean. A horse can hardly have too deep and broad a chest, too strait a back from withers to croop, or too thin and delicate a neck near the head. Under the cut above we give the names of the various parts, and some of the blemishes and diseases of the parts. Many diseases are only known in fancy, and almost all are known by several different names. (1862)

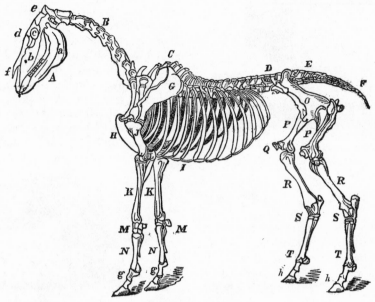

Skeleton of a Horse.

A. The Head.—*a* The posterior maxillary or under jaw.—*b* The superior maxillary or upper jaw.—*c* The orbit, or cavity containing the eye —*d* The nasal bones, or bones of the nose.—*e* The suture dividing the parietal bones below from the occipital bones above.—*f* The inferior maxillary bone.

B The Seven Cervical Vertebræ, or bones of the neck.—C The Eighteen Dorsal Vertebræ, or bones of the back.—D The Six Lumbar Vertebræ, or bones of the loins.—E The Five Sacral Vertebræ, or bones of the haunch.—F The Caudal Vertebræ, or bones of the tail. generally about fifteen.— G The Scapula, or shoulder-blade.—H The Sternum, or fore-part of the chest.—I The Costæ or ribs, seven or eight articulating with the sternum, and called the *true ribs ;* and ten or eleven united together by cartilage, called the *false ribs.*—J The Humerus, or upper bone of the arm.—K The Radius, or upper bone of the arm.—L The Ulna, or elbow. The point of the elbow is called the Olecranon.—M The Carpus, or knee, consisting of seven bones.

N The Metacarpal Bones. The larger metacarpal or cannon or shank in front, and the smaller metacarpal or splint bone behind.—*g* The fore pastern and foot, consisting of the Os Suffraginis, or the upper and larger pastern bone, with the sesamoid bones behind, articulating with the cannon and greater pastern, the Os Coronæ, or lesser pastern ; the Os Pedis, or coffin-bone ; and the Os Naviculare, or navicular, or shuttle-bone, not seen, and articulating with the smaller pastern and coffin bones.—*h* The corresponding bones of the hind-feet.

O The Haunch, consisting of three portions : the Ilium. the Ischium, and the Pubis.—P The Femur, or thigh.—Q the stiff joint with the Patella.—R The Tibia, or proper leg bone ; behind is a small bone called the fibula.—S The Tarsus. or hock, composed of six bones. The prominent part is the Os Calcis, or point of the Hock.—T The Metatarsals of the hind leg.

Keeping One Horse

To keep a horse, so that he may render the longest continued, and greatest amount of valuable service, it is essential that he should be provided with comfortable quarters.

The stable must be dry, warm, light, and with good ventilation; the latter to be managed on some plan that will not cause currents of air to pass over the inmate. A loose box is a great comfort to a horse. The term "box" seems to frighten most people. So far from being some complex arrangement, only within reach of the wealthy, it is as simple as possible. A reasonably wide stall may be used as one: but a few inches more width is desirable, depending on the size of the animal. If hard-worked, a horse rests more completely in a loose box than in a stall, and, when idle, he can exercise himself in it sufficiently to prevent stiffness or swelling of the legs.

A good bed keeps a horse clean when lying down, and aids his rest. Wheat straw is the best article of which to make it, and about five cwt. per annum will be needed. (1881)

The Future of the Horse

When the agricultural world was invaded by labor-saving machinery, very much of which was operated by the labor of horses or cattle, the cry arose that men were to be crowded out of employment. The same cry was raised in a multitude of manufacturing industries whose operatives declaimed against each new appearance of machinery perfected in the line of economy of human labor. But results have proved the groundlessness of such fears. The introduction of improved machinery simply enlarged production, enabling it to keep pace with a rapidly increasing demand, while men's labors were found to be as much in demand as before, but under somewhat different conditions.

In like manner, when a few years ago the possibilities of electricity began to become apparent, and the equipment of horsecar lines with this new motive power was seen to be a coming event, the cry was raised that the breeding of horses was to receive a serious check. Later came the safety bicycle, and thousands, who before depended upon horses for locomotion, sold their teams, and took to the less expensive silent steed. Here, then,

was another evidence of the passing of the horse. Just how far these fears are likely to be realized cannot yet be fully known, for the application of electricity as a motive power has not yet become at all universal, nor has the bicycle attained the full measure of its popularity, distinctly and widely popular as it has already become ; but this much is certain, the time is not far distant when tens of thousands of horses than now draw tram and street cars will be replaced by the trolley or storage battery system.

In the continuous discovery of electrical possiblities it is impossible now to say whether electricity may not replace horses in other lines of work, such as teaming between points joined by electric railroads, on some of which even now freight, as well as passenger cars, are being run, the freight service being employed in the night when the track is clear. All these signs point to a gradual encroachment upon the work of light and heavy draft horses by the electric current, whose wonderful powers are being so rapidly demonstrated. One need not be an alarmist to foresee a lessened demand for work horses from that demand that would undeniably have existed had electricity not entered this field.

Draft horses will always be needed, but it is the part of wisdom to bear in mind a very possible curtailment of their usefulness, and, therefore, their value, in the future. The wise man lays his plans for possible contingencies. If he breeds horses of a particular grade, and thinks he sees a lessening demand for his product in the future, he will straightway get himself in readiness to meet changed conditions. Will he give up breeding horses? Not necessarily. He will simply find out what class of horses is likely to be in demand, electricity or no electricity, and will endeavor to meet the requirements of a changing market.

The person who looks carefully into the conditions that exist in society at the present time cannot fail to be impressed with the rapid accumulation of wealth on every hand, and the tendency to get as much comfort and pleasure out of it as possible. One marked result of this is an enormously increased demand for stylish and high-spirited driving horses, matched pairs, and fast roadsters. The noted horse-breeding portions of the country are being constantly searched by the agents of wealthy men looking for horses that meet these conditions, and when they find what they want, the matter of price rarely stands in the way of a purchase. The state of Maine, for instance, is being constantly traversed by these men, and scores of horses are being as constantly shipped out of the state, leaving behind them large sums of money as their equivalent. It is the same in other regions where special attention has been given to the raising of fast and stylish roadsters. Nothing of an economic or industrial nature comes into this demand, and changing conditions of life are not at all likely to affect it.

There are many sections of our country excellently adapted to the raising of first-class horses, which have as yet, not been developed. A substantial beginning in this direction need not necessarily involve the investment of a large amount of capital. The main outlay would be for a pure-bred sire of a strain desired to perpetuate. From this start-ing point, the character of the stud should be constantly improved by the infusion of better blood. I have never been an advocate of an attempt on the part of the average farmer to raise fast horses, nor do I now advocate it; but I think the time has come to look the mat-

THE CLYDESDALE STALLION "NUBIAN."

ter squarely in the face. If farmers are to continue to breed horses, it is the part of wisdom to let such breeding run in lines where there is the greatest demand and the most money. The secret, or, at least one of the secrets, of successful farming is to find out what the public wants and then to furnish the very best quality of the article desired. If the public wants a particular type of horse and is willing to pay liberally for good specimens of this type, and if farmers are to continue to breed horses, then it is certainly wise to supply the demand. (1893)

Raising Ducks

Two years' experience with a flock of Pekin ducks has convinced the writer that there is a satisfactory profit in raising these birds. But the conditions must be favorable, and these include a water-run, either a stream or pond, in which the ducks can gather food, and a house conveniently arranged for securing the eggs.

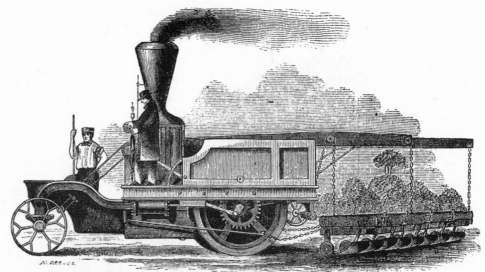

FAWKES' AMERICAN LOCOMOTIVE STEAM PLOW.

The first year a beginning was made with a trio of the birds, and these were conveniently accommodated in an ornamental rock-work house among some evergreens on a side lawn. A neighboring stream of water kept the ducks out of mischief in the day time, and they came home regularly at night; they were not let out in the morning until the eggs had been secured. The two ducks laid 202 eggs the first season; of these, some were sold, some eaten, and the remainder set under hens. Bad luck, in which may be included the destruction of three-fourths of the eggs when under the hens, and the killing of several of the sitters by a stroke of lightning, which went through the poultry house, reduced the produce of young ducks to between 30 and 40. But one of these died after leaving the nests, which goes to prove the hardiness of this variety. The young ducks thrived well, and when six to eight months old weighed five to six pounds on the average. A large por-

tion were killed and eaten; the flesh was found to be remarkably juicy and tender.

The second year it was necessary to provide larger accommodations, and a house was made for them on the bank of a pond adjoining a brook in which there are abundance of watercresses and other food, both vegetable and animal. The watercress is eaten with avidity by ducks, and has myriads of snails and other water animals upon it.

A plan of this house is shown at figs. 1 and 2. For 50 to 100 ducks it should be 30 feet long, 12 feet wide, and from 4 feet high at the front to 6 or 8 feet in the rear. Entrance doors are made in the front, which should have a few small windows. At the rear are the nests; these are boxes open at the front. Behind each nest is a small door through which the eggs may be taken. It is necessary to keep the ducks shut up in the morning until they have laid their eggs, and a strip of wire netting will be required to enclose a narrow yard in front of the house. Twine netting should not be used, as the ducks put their heads through the meshes and twist the twine about their necks, often so effectively as to strangle themselves. To avoid all danger, the wire fence should have a 3 or 4-inch mesh.

Fig. 2.—GROUND PLAN OF THE HOUSE.

The Pekin ducks are prolific layers; a fair yearly product for a duck in its second year is 120 eggs, and 60 to 80 for a yearling. Their feathers are of the best quality, white, with a creamy shade, and 5 ducks weighing 5 pounds each, have yielded, killed in the winter time when fully feathered, more than one pound in all. It will be right to pick the ducks when moulting is beginning; the feathers are then loose and are picked easily, and without injury. This will considerably increase the yield of feathers, and will prevent a useless loss; otherwise the loose feathers from 20 ducks will be found spread over their whole range. (1864)

THE SHEEP-FOLD.

" Now sacred Pales, in a lofty strain
I sing the rural honors of thy reign.
First, with assiduous care from Winter keep,
Well foddered in the stalls, thy tender sheep."
DRYDEN'S VIRGIL.

Care of Sheep and Lambs

Something more than high prices for wool and mutton is needed to make sheep raising profitable. They must be well cared for, particularly at the lambing season. If the ewes have been judiciously fed, neither stinted nor pampered, the labors of the keeper will be greatly lightened. The lambs will come into the world, vigorous and active, requiring little more than the care afforded by the ewe.

A few points will always need attention. Experience is the best teacher, but many have their first flock of ewes to manage, and to such, the following practical suggestions will be timely. From the first, pains should be taken to render the flock tractable. A few handfuls of oats or corn scattered among them on each visit will make the master always welcome. In this way a flock may soon learn to be led to any desired place. Pregnant ewes and their progeny are often injured by their efforts to escape when being driven to or from an enclosure. If the weather be clear and mild, it is preferable to have lambs dropped in the pasture. The field for their accomodation should be dry, and free from ditches or sunken spots. But during

Fig. 1.—VIEW OF A CONVENIENT DUCK-HOUSE.

Hamburg Fowls.

A PAIR OF AYLESBURY DUCKS.

SHEEP TENDING....From a Painting by H. Brittan Willis, London.—*Engraved for the American Agriculturist.*

cold nights and in rainy weather, shelter is essential. Make the shed for the ewes roomy, and allow plenty of ventilation. Where the flock is large, the shed should be divided into temporary pens to accommodate not more than 20 or 30 head. In the moving about and confusion of a larger number, the young mother may be crowded away from her offspring, and the lamb be unable to suckle.

The first great point to gain is that the young should early get a good supply of food from the dam. It needs this both for nourishment, and for the medicinal effect which the first drawn milk has on the digestive organs. If mechanical assistance be needed, let it be of the gentlest character, and only in conjunction with the efforts of the animal.

Where young lambs are found astray without a natural protector in the flock, if no foster mother can be provided, they may be given to the children to bring up as cossets. Feed them with warm fresh cow's milk. They will readily learn to drink it by giving them a quill with a strip of cloth tied around it to suck through at first. Sheep reared in this way at the house are likely to have extra care, and they usually bring an extra price in market, besides giving much pleasure to the young members of the family while rearing them.

Abundant nourishment should be provided for lambs in the flock, by giving good pasture to the ewes. If grass be short, a daily small allowance of oats or corn will pay both in the lambs and the fleece. (1863)

Shelter for Sheep

The sheep is an animal which will endure much exposure, and its health is better when it is exposed to most of the natural changes of the atmosphere, but not to storms or wet ground. The Merino is probably the hardiest of those breeds which are most profitably bred in this country, and will bear most exposure.

There are many fine flocks, healthy and vigorous, particularly in the prairie states, which never have more shelter than is afford by a board fence, or an Osage orange hedge. For all this, sheds would be a great comfort both to the sheep, and to the shepherds. The straw and rail shelters first, to be succeeded by more substantial sheds, and these again by good sheep barns — this is the order of progress.

A sheep barn must afford shelter both for the sheep and their feed — hay, straw and grain. It should be supplied with flowing water or a good well. The site should be perfectly dry and sheltered from winds to avoid much drifting of snow, and it is best to have it large enough to accomodate the entire flock. Or, if it is impracticable to have all in one barn, then the barns should be near together, and if possible, placed so as to afford more effectual shelter.

The barn should always be built on the side-hill principle, even though it be on level ground — the sheep rooms being on the lower floor, and the entire space above being used for hay and grain. Where there is no available hill, the sheep floor may be depressed a little, and causeway raised so that teams may be driven in upon the main floor.

For large flocks a convenient arrangement is a main building with low wings, which indeed are only closed sheds. The number of sheep which may be accomodated in a certain space, varies with each different breed. The larger families of Merinos need more space, and the South-Downs and Long-Wools still more (near twice as much as the first names). It is better to have too much space than too little, at any rate.

Double-feeding racks may be so arranged as to form partitions, to subdivide the 100 or 150 or however many, but they will not be a sufficient separation for the chief divisions of a large flock. The rooms and yards should be entirely distinct; and it is very desirable also to have a detached shed wherein to place any part of the flock which may be diseased, or which may have been exposed to disease.

In all buildings for sheep, the floors above them should be perfectly tight (tongued and grooved) to prevent the sifting through of hay

seed and dirt, and all hay racks used should be so constructed that dust will not get into them. Sufficient litter should also be used, to prevent injury to both sheep and fleece by the floor. (1864)

Sheep Troughs

One of the annoyances connected with feeding grain to sheep and other stock is the liability of the trough to become dirty, when it must either be upset, or cleaned out with a bit of board, broom, or wisp of coarse fodder, as most convenient. Also troughs that are from necessity left out doors during winter, become partly filled with snow many times each week, consuming a great deal of time to clean them. To avoid all this trouble I have planned the following self-cleaning troughs that are cheap and easily made.

The simplest plan is shown in fig. 1; the

Fig. 1.—A REMOVABLE FEED TROUGH.

THE CHILDREN AND THEIR PET LAMB.—*Engraved for the American Agriculturist.*

trough is hung at each end on wooden pins, *t*, *t*, attached to the plank-post. The ends of the trough project upward about ten inches; near the upper end of each is a half-inch hole, that corresponds with one made in the post; by passing a loosely fitting pin through these holes the trough is firmly held in position. During stormy weather or at any time, by turning the trough bottom up, as in fig. 2,

Fig. 2.—TROUGH INVERTED.

and inserting the pin at *n*, the trough is cleaned from all dirt, kept dry and free from rain, snow, or ice. Posts may be driven in the ground or connected at bottom by a scantling as desired. With the trough shown in fig. 3,

Fig. 3.—HINGED FORM.

pins are dispensed with. It is constructed by cutting a V-shaped notch in the top of the plank post, in which rests the end of the trough. An iron hinge, *e*, at each end connects the top of the post and trough. At any time, simply raise the opposite side of the trough, and fold it over against the side of the post as shown in fig. 4. A trough in this position is, for all practical purposes, just as good as though bottom up; however by leaving the post nearly as wide again at top,

and on a level with the top of trough when right side up, the trough can be turned bottom up, when properly hinged. This trough is also easily made portable by connecting the ends below the trough. (1882)

4.—TROUGH FOLDED OVER.

FAMILY PETS—PARTING OF FRIENDS.

AUTUMN SHOOTING. *Drawn and Engraved for the American Agriculturist.*

Some Good Breeds of Dogs

There is no race of animals which shows a greater diversity of qualities than the canine. Between the prowling coyote and the noble and intelligent St. Bernard there is a wider range of difference than in any other genus of the brute creation. These differences are the results of domestication, selection, and training. Whoever would doom the entire race of dogs to a sweeping destruction would deprive the human family of the most intelligent, affectionate, and interesting of its dumb servants and pets.

The annual bench shows in New York bring out the best specimens of several breeds, from the largest to one of the most diminutive. There is the Great Dane, the tallest and hugest of all. The origin of this breed is not clearly known, but it possesses characteristics of both the mastiff and hound. A smaller dog is a fox terrier, a breed formerly employed in fox hunting, to unearth the wily game from its hole, but kept chiefly, in recent years, as household and personal pets. The fox terrier is a smooth-haired, clean-cut animal, the weight not exceeding sixteen pounds.

Another terrier is the Dandie Dinmont terrier, a breed having descended from the two dogs raised by James Davidson, the original of Scott's "Dandie Dinmont." These dogs are low on the ground, with heads like an otter hound's, and long bodies covered with shaggy coats of the famous pepper and mustard colors. The Dandie Dinmont weighs from 18 to 24 pounds, and is one of the most intelligent and plucky dogs in existence. It is eager in pursuit of game, and will boldly attack its prey and fight to the death.

The Irish water spaniel has long silky ears, a decided top-knot, the entire body covered with fine curly hair, and a "spike" tail. These dogs are eager in retrieving game, leaping fearlessly from a bank 25 feet or more into the water to recover. They swim and dive with the greatest east and zest, and are valuable dogs for hunting water fowls.

The English setter is one of the most widely known and favorite breeds of dogs kept in this country. In the prairie regions and remote West, setters are found especially valuable in hunting grouse and other game.

The soft, fine coat, silky ears, feathered legs and tail, and general appearance of these dogs is familiar to every one. (1889)

The Old English Sheep Dogs

Many of our readers doubltess suppose that the Scotch colley is the only "sheep" dog. To such it may be a surprise to learn that there is another breed very different from the colley that is fully as much of a sheep dog. We allude to the old English bobtail sheep dog, generally born without a tail, or at most a very short one, and with a rough, shaggy coat of hair.

They are a much heavier, more compact dog than the colley, and although without his beauty, they may safely challenge competition for usefulness; in fact, those who have them say they are much better natural workers than the colley. One, owned in Western Pennsylvania, although only a year old, will stop the rams from fighting, and "lay by" a weak sheep all day, keeping other stock from running over it; will run near his master, or where he thinks he can be heard, barking like all possessed if a sheep gets into any trouble, such as getting fast in a wire fence; will guard anything of his master's he finds in the field, and when a sheep is brought home, killed and hung up for dog meat, "Bob" will establish himself as guard, and not a dog dare go nearer than the distance "Bob" prescribes.

All this is without any training, being simply his natural disposition. This natural disposition is easily accounted for, as they are too odd-looking to become general favorites on the show bench, and therefore have been bred altogether for use. (1886)

THE STRAW-YARD AT CHRISTMAS—From a Sketch by Herring.
(Engraved for the American Agriculturist.)

THE "RED" BERKSHIRE SOW "BELLE."—Engraved for the American Agriculturist.

121

IX   Beekeeping

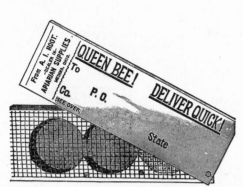

THE BENTON SHIPPING AND MAILING CAGE.

FOUNTAIN FOR GIVING BEES ACCESS TO WATER.

Queens

The most important personage in the hive is the queen, or mother-bee. She is called the mother-bee because she is, in reality, the mother of all the bees in the hive.

THE QUEEN AND HER RETINUE.

If you deprive a colony of their queen, the bees will set to work and raise another, so long as they have any worker-larvae in the hive with which to do it. This is the rule, but there are some exceptions: the exceptions are so few, however, that it is safe to assume that a queen of some kind is present in the hive, whenever they refuse to start queen-cells from larvae of a proper age. (1895)

Drones

These are large, noisy bees that do a great amount of buzzing, but never sting anybody, for the very good reason that they have no sting. The beekeeper who has learned to recognize them both by sight and sound, never pays any attention to their noise, but visitors are many times sadly frightened by their loud buzzing.

The body of a drone is hardly as long as that of a queen, but he is so much thicker

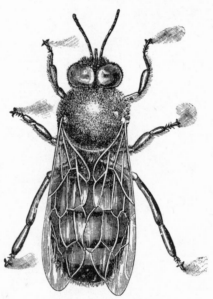

DRONE-BEE.

through than either queen or worker, that you will never mistake him for either. He has no baskets on his legs in which to carry pollen and his tongue is so unsuited to the gathering of honey from flowers, that he would starve to death in the midst of a clover field.

We have no satisfactory evidence that the meeting between queens and drones takes place not very high up from the ground. Several observers have reported seeing this meeting not very far from the hives, during the swarming season. The queens and drones both sally forth during the middle of the day, or afternoon, and in from fifteen minutes to an hour, or possibly a couple of hours, the queen returns with a white appendage attached to the extremity of her body, that microscopic examination shows to be the generative organs of the drone. (1890)

Swarming of Bees

All animated nature seems to have some means of reproducing its like, that the species may not become extinct; and, especially among the insect tribes, we find a great diversity of ways and means of accomplishing this object. In the microscopic world we find simple forms of animal life contracting themselves in the middle until they break in two, and then each separate part, after a time, breaks in two, and so on. With bees we have a somewhat similar phenomenon. When a colony gets excessively strong, the inmates of the hive, by a sort of pre-concerted, mutual

W. F. CLARKE'S SWARMING-DEVICE.

A SMALL STARVED-OUT SWARM.

WHITE CLOVER.

The most important is the common white clover *(Trifolium repens)*, which everybody knows is perhaps at the head of the entire list of honey-producing plants. We could better spare any of the rest, than our white clover that grows so plentifully as to be almost unnoticed almost everywhere. But little effort has been made to raise it from the seed, because of the difficulty of collecting and saving it.

The common red clover—*T. pratense*—yields honey largely some seasons, but not as generally as does the white, nor do the bees work on it for as long a period. While working on red clover, the bees bring in small loads of a peculiar dark-green pollen; and by observing this we can usually tell when they are bringing in red-clover honey. (1895)

Basswood

With perhaps the single exception of white clover, the basswood, or linden, as it is often called, furnishes more honey than any other one plant or tree known. It is true, that it

does not yield honey every season, but what plant or tree does? It occasionally gives us such an immense flood of honey that we can afford to wait a season or two, if need be, rather than depend on sources that yield more regularly, yet in much smaller amounts. If a beekeeper is content to wait—say ten or fifteen years for the realization of his hopes; or if he has an interest in providing for the beekeepers of a future generation, it will pay him to plant basswoods.

agreement, divide themselves off into two parties, one party remaining in the old hive, and the other starting out to seek their fortunes elsewhere—they begin swarming.

If you can contract the size of the hive when honey is coming in bountifully, the bees will be very apt to take measures toward swarming, about as soon as the combs are full of brood, eggs, pollen, and honey. They will often wait several days after the hive is seemingly full, and this course may not cause them to swarm at all, but it is very likely to. As soon as it has been decided that the hive is too small, and that there is no feasible place for storing an extra supply of honey where it can be procured in the winter, when needed, they generally commence queen cells. Before doing this I have known them to go so far as to store their honey outside on the portico, or even underneath the hive, thus indicating most clearly their wants in the shape of extra space for their store, where they could protect them.

Want of room is probably the most general cause of swarming, although it is not the only cause; for bees often swarm incessantly when they have a hive only partly filled with comb. Yet, there is much truth in the old adage—

"A swarm of bees in May
Is worth a load of hay;
A swarm of bees in June
Is worth a silver spoon;
A swarm of bees in July
Is not worth a fly."

(1895)

Clover

While most persons seem to tire, in time, of almost any one kind of honey, that from the clovers seems to "wear" like bread, butter, and potatoes; for it is the great staple in the markets; and where one can recommend his honey as being pure white clover, he has said about all he can for it.

The cut will enable any one to distinguish at once the basswood when seen. The clusters of little balls with their peculiar leaf attached to the "seed-stems" are to be seen hanging from the branches the greater part of the summer, and the appearance, both before and after blossoming, is pretty much the same. The blossoms are small, of a light, yellow color, and rather pretty; the honey is secreted in the inner side of the thick fleshy petals. When it is profuse it will sparkle like dewdrops if a cluster of blossoms is held up to the sunlight. (1895)

Bee Pasture in August

In localities where there are abundant fields of buckwheat the bees appear to work just as industriously as they did among the basswood blossoms, the rich supply from which has now ceased. But besides buckwheat, the many golden rods and other members of the same family, the *composites* keep in flower until hard frosts, and afford an appreciable supply as long as it is safe for the bees to gather it. The honey from buckwheat has a flavor which to some is unpleasant, while other prefer it to honey from any other source. The honey from the many *Solidagos*, of which some species are found on high lands, and others in swamps, in great profusion, are a source of honey which has not heretofore been generally appreicated. Prof. A. J. Cook, who has devoted much attention to honey-producing plants, says of the golden rods: "They yield an abundance of rich golden honey, with a flavor that is not surpassed by any other." (1889)

Wintering Bees

The profits of beekeeping are probably more directly affected by the strength and health of the colonies in the spring, than by any other cause. To secure a good start, we must of course get our bees well through the winter. It is not a question of a few pounds of honey, or of syrup, more or less, but really a question of health. If Mr. Root's bees, when the basswood trees are in full bloom, can store 14 lbs. of honey per hive on an average work-day, when he had 40 colonies, other people's bees may perhaps do equally well; at any rate, healthy bees with abundant pasturage, earn for themselves the very best winter treatment we can give them.

Prof. Cook's system, which has proved very succuesful, is about as follows: He secures strong colonies in the autumn. When the bees cease to store honey, and have about 30 lbs. of honey in the frames, the brood contained in the frames is allowed to hatch, and the frames are then removed, while the full frames are set in towards the center, and par-

tition boards are set on each side. The side spaces are packed with chaff or other non-conducting material, a sheet of factory cloth (common sheeting), is placed over the six or eight frames left in, first laying a short stick over the tops of the frames, so that the bees can pass over the tops if they wish to. Over all a sack of burlap, filled with dry sawdust is laid. And when severe weather threatens, the bees are removed to a cold cellar, where a uniform temperature of 40 to 45 degrees is maintained throughout the winter. Here the hive covers are removed and the entrance holes are opened.

Experiments have demonstrated, that at the temperature named, the bees will consume very little honey, and are less liable to diarrhea than if either warmer or colder. In any case, they will be under the necessity of emitting their excrements, which from their neat habits they dread to do; unless they can fly out, they are liable to the disorder named.

MILLER'S ROPE CARRIER.

Hence, they must be watched, and if there appear evidences of diarrhea, the hives ought to be removed from the cellar on the first bright, warm day, placed each upon its old stand, if possible, and after the bees have had a flight, and voided their excrements, they should be returned to the cellar. They should not be removed finally from the cellar before they can go to work. (1886)

Wintering Bees in a House

The method of wintering bees practiced by Mr. Hogan of DuPage Co., Ill., is thus described: He builds a house of suitable size to contain his stocks, something like an ice house, of joists, clapboarding the outside and lining the inside with matched siding, leaving a space of four inches all around. This is filled with chaff, and the hives are arranged four tiers high all around the inside. To ventilate it, he constructs an air tube from the

outside under ground to the center of the house, where it is admitted through a perforated board or plate of metal. At the top a passage is made for the heated air to escape. The whole is arranged to exclude every particle of light. The hives are left open as in the summer. The heat generated by the bees is sufficient to keep the air warm enough for their safety and comfort. (1862)

Honey Boxes

Several recent readers have asked for a description of what we have called "the simple unpatented glass honey boxes used by Mr. Quinby." The annexed engraving the book received from Mr. Q. will show the general form. It is, say 5 inches wide, 6 inches long, and 5½ inches high. The top and bottom are made of boards ¼ inch thick. The four corner uprights are square pieces say 5/8ths of an inch through and 5 inches long. These are set up at the corners and held in place by small nails driven into their ends, through the bottom and top board.

For the sides, common window glass is cut into the required size. It can be cut without waste, by choosing the panes of the right size. The glass pieces are placed against the uprights, on the outside, to complete the box, and are held there by bits of tin, slit with shears part way through the middle, and the slit end shoved through the uprights, corner-wise, from the inside outward. The slit end is then bent over the edges of the glass, one part

to the right, the other to the left. The box is here represented bottom side up to show the hole in the bottom piece. These boxes are set side by side over a common box hive, the holes meeting other holes in the top of the hive. A box cover is set over the whole tightly, to shut out rain and keep the boxes dark, otherwise the bees will not work in them. (1859)

The very nicest veil is one made entirely of silk tulle, although it is somewhat more expensive. The material is so fine that a whole veil of it may be folded up in a small vest pocket. It neither obstructs the vision nor prevents the free circulation of air on hot days. A cheaper one is made of grenadine with a facing of silk tulle not sewn in.

Some boys use with satisfaction what is called the Hopatcong. It is a hat worn in India and other hot countries, and is slowly working its way into this country, particularly in the South. It is made of palm leaf, and it is supported above the head in the manner illustrated. (1895)

MRS. R. H. HOLMES' BEE HAT.

Bee Stings

It is true that bees can not bite and kick like horses, nor can they hook like cattle; but most people, after having had an experience with bee stings for the first time, are inclined to think they would rather be bitten, kicked, and hooked, all together than risk a repetition of that keen and exquisite anguish which one feels as he receives the full contents of the poison-bag, from a vigorous hybrid, during the height of the honey season.

I would pull the sting out as quickly as possible, and I would take it out in such a way as to avoid, as much as possible, squeezing the contents of the poison bag into the wound. If you pick the sting out with the thumb and finger in the way that comes natural, you will probably get a fresh dose of poison in the act, and this will sometimes prove the most painful of the whole operation, and cause the sting to swell when it otherwise would not have done so. The blade of a knife, if one is handy, may be slid under the poison bag, and the sting lifted out without pressing a particle more of the poison into the wound. When a knife blade is not handy, I would push the sting out with the thumb or finger nail in much the same way. (1895)

Veils

The necessity of using face protections will depend very largely upon the race of bees to be handled. Its use will, in any case, give the apiarist a sense of security that will enable him to work to much better advantage than he would if continually in fear of every cross bee that chanced to buzz near his eyes.

HOPATCONG HAT AND VEIL.

J. H. MARTIN'S BEE-SUIT.

BEE-HATS FOR WOMEN.

Bee's Wax

Whether bees make honey, or simply collect it, may be a subject of discussion; but we believe there is no question in regard to wax, for bees assuredly do make it.

Wax from the hives varies greatly in hardness. Some specimens are so soft that it seems as if they could not stand the weight of bees at all, when made into foundation, while others are so hard that it is difficult to roll them at ordinary temperatures. The wax of commerce, when it is bought in quantities, is composed of cakes of all sizes and of all colors, from nearly white to nearly black, the intermediate shades comprising almost all the colors of the rainbow. (1895)

Hive Making

Unless you are so situated that freights are high, and unless, also, you are a mechanic, or a natural genius in making things, you had better let hive making alone. Hives can be bought, usually, with freight added, for a great deal less, than the average beekeeper can make them himself. But there is lots of fun in making things, even if they are not so well made; and there are some rainy or wintry days in the year, when, if you are a farmer, for instance, you can as well as not, and at little or no expense for time, make a few hives and other fixin's.

While it is very important to have good well-made hives for the bees, I would by no means encourage the idea, that the hive is going to insure the crop of honey. I think, as Mr. Gallup used to say, that a good swarm of bees would store almost as much honey in a half-barrel or nail keg, as in the most elaborate and expensive hive made, other things being equal.

A good hive must fill two requirements reasonably well to be worthy of that name. 1. It must be a good home for the bees; 2. It must in addition be so constructed as to be

127

convenient to perform the various operations required by modern beekeeping. The first of these requirements is filled very well by a good box or straw hive. Bees will store as much honey in these hives as in any, and in the North they will winter and spring as well in a straw hive as in any other. They do not, however, fill the second requirement; and to meet this, the movable frame hive was invented. (1895)

A New Home-Made Beehive

The engraving represents the section of a beehive made and used by M. S. Woodford of Erie Co., N.Y., who thus writes: "I propose to describe for the benefit of the readers a beehive that I make and have used for 4 years. First I make the four sides of a

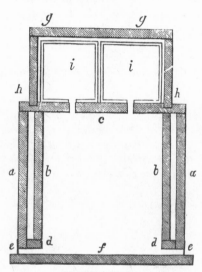

HOME-MADE BEEHIVE.

box, *b*, as recommended by Mr. Quinby, 12 inches inside each way. I then make another, *a*, 15 inches inside each way, and one inch deeper than the first one. A top board, *c*, is made, large enough for the larger box, and rabbited back from the edge all around to receive the cap, *g*, made 13 inches inside. This top I nail to both boxes one within the other. The strips, *d*, are then nailed on the inside of the outer and on the lower end of the inner box. This closes up the space between the two, and makes a hive warm in winter and cool in summer.

The bees will not come out of such a hive in winter, unless the weather is warm enough to allow them do so without harm. To prevent millers depositing their eggs under the edge of a hive I take a strip of hoop iron, *e*, 1½ inches wide, and nail around the bottom, forming a band. This may be readily done by any one not a mechanic, by cutting it in four pieces, one to fit each side and make close joints at the corners, punching holes

and filing the edge straight. Nail them on the hive so that the iron will project about half an inch below. The edge or iron will rest on the bottom board, *f*, leaving a half inch space between the bottom of the hive and the board, giving the bees a chance to operate on the board out to the iron. To give the bees an entrance, a notch is cut in the front piece of iron 3/8ths of an inch deep and 3/4th of an inch long. I also give them another entrance, three and a half inches above the alighting board, by inserting a plug which reaches through the two boxes with a half inch hole in the center. Some may think this hive too expensive, but from four years experience I have found that it pays. (1864)

Cheap and Good Straw Hives

E. J. Ferris, of Lake Co., O., J. T. Smith, of Uniontown, and several others, inquire how to make the straw hives referred to in the July issue. While at M. Quinby's, we examined quite a variety of straw hives, mostly patented by different parties. We will describe one of the best forms, one of which is

unpatented, and can be made by any person with moderate skill.

The size depends upon what is required. If for a particular kind of honey boxes or movable frames, the size must be made to correspond with what is wanted. It is a square or parallelogram, to be covered with a flat board to receive the surplus boxes, and

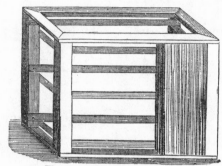

Fig. 1.—FRAME FOR THE STRAW.

over this a wooden box with sloping or flat roof, and projecting over the sides to shed rain. The essential part of body of the hive is made as follows:

For the upright corner pieces, cut 2-inch square stuff to the required length. Upon the inside of these nail three pieces of lath for the

THE VINEYARD APIARY, AND "SWARMING" THE GRAPEVINES.

sides and ends, putting one strip around both top and bottom, and one in the middle, as seen in fig. 1. Then nail flat thin strips, 2 inches wide, around the top and bottom, covering the ends of the uprights, as also shown in fig. 1. Next cut clean, straight straw, in a cutting box, to just the required length to fit into the sides. Pack this straw in

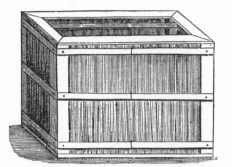

Fig. 2.—THE STRAW FRAME COMPLETED.

firmly upon the sides, and nail strips of lath on the outside, as shown in fig. 2, and the thing is done. To prevent crowding off the inside strips while packing in the straw, it is well to have a false box just the size of the inside, and slip this in while packing the straw. To prevent the spreading of the lath, bind

them together at the middle points with wire running through the straw, especially on the longer sides of the hive. Straw hives are grateful to bees, cool in summer and warm in winter, and with the straw standing perpendicular, as above described, it sheds off all rain. As 2 inches thickness of straw would seem to be more than is needed, if the corner pieces be 2-inch stuff the outside slats might be let into the pieces the depth of their thickness, though this would somewhat increase the labor of making them. They are quickly and cheaply made, and are neat in appearance, especially if the wood be planed; this is not essential, however. If the woodwork be painted, they will look still more attractive. (1863)

AN EXPERIENCE THAT "BLESSED BEES" DIDN'T TELL OF.

WORKER.

STILLMAN-CH.
OUR ORIGINAL HOUSE-APIARY.

THE LAWN OR CHAFF-HIVE APIARY.

X   Fruits & Berries

APPLE GATHERING—From a Painting by Jerome Thompson.

Fruit Picking & Fruit Pickers

The choicest specimens of pears and apples often grow on the ends of long, slender branches, which will not support a ladder, nor a man while plucking the fruit. When long ladders are leaned against the outsides of trees, many of the small limbs and fruit-buds are broken off. Sometimes pear trees grow so tall that the limbs are not strong enough to bear a small boy in the tree, nor on a ladder resting against it, unless it is supported with guy ropes.

An orchard ladder should have its lower ends shod with iron, in the form of a wedge, to enter the ground readily, and to hold the lower end when putting it up and down; see fig. 2. Set the ladder nearly perpendicular, and stay it with two guy ropes from the top of the ladder fastened to trees or stakes, or fences, as shown in fig. 1. The ropes need not be larger than a common clothes line. A man can ascend to the very top of a long ladder secured in this way, and pluck half a bushel or more of fruit with entire safety.

A large bag suspended on one shoulder, and under the arm on the opposite side, is much more convenient than a basket, as there is no danger of letting the fruit drop, as

with a basket; and both hands are always free, whether the picker be in the tree or on a ladder. if fruit be borne upon long, slender branches, by drawing the ends inward or downward, as represented in the illustration,

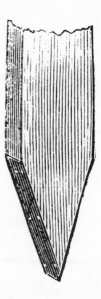

all may be reached for several feet on each side of the ladder.

A fruit picker having a handle 4 or 5 feet long, will sometimes be found convenient for taking that which can not be reached otherwise. The number of styles of fruit pickers is great, and much ingenuity has been expended on more or less complicated contrivances. With this as with most other im-

132

plements, the simplest is the best. We give a figure of one, fig. 3, which can be easily made, and which will accomplish the purpose as well as those which have a great deal of machinery about them. A stiff wire is bent in the form here shown, to which a bag is attached, and the whole is fastened to a handle, which may have a hook near the picker or on the other end. In bending the wire, the lip or projection to the ring should be made so narrow that a small apple can not slip through. A similar lip is formed by the manner in which the ends of the wire are fastened, one on each side of the handle. With this picker, the ring is put over the apple, and by drawing or pushing the stem passes into one of the lips and the fruit falls into the bag. (1865)

A Convenient Fruit Picker

A leading daily paper, a short time ago, announced that some one had invented an implement for picking fruit, and went on to describe one of the oldest devices for the purpose, as if it were a new invention, and it was no doubt new to that writer. Fruit pickers have long been in use, and the records of the Patent Office will show that a great variety have been invented. Some of them are too complicated to come into general use. A fruit picker is only of use to bring down such fruits as cannot be reached by hand, from a ladder or otherwise, as those that grow in such

A FRUIT PICKER.

positions are usually the finest specimens, and too valuable to be shaken off, it is worthwhile to take some trouble to secure them in a perfect condition.

We have shown in a former volume that a fruit picker may be made from an old fruit can; the can, with a V-shaped notch cut in the upper edge, is fastened to a suitable pole, putting a lock of hay, some paper, or other soft material in the can to prevent bruising. By catching the stem in the V-shaped notch, the fruit is readily detached and brought down. But such a picker, though effective, is only a make-shift, and as it has to be lowered to remove each fruit, it makes hard work. A better picker, and one that is quite as good as any, is one invented by Jas. H. Ten Eyck, Auburn, N.Y., who sent the sample from which the engraving was made.

It is a bottomless dish, 3 inches high, 5 inches across at the top, and 3½ inches at bottom, made of tin, with a socket to receive a pole of convenient length. Above the edge there project two stout iron wires of the shape seen in the cut; these are close together below and diverge above, and serve as fingers to catch the stem and detach the fruit. The most important portion of the affair is not shown in the engraving; this is a bottomless bag, or hose, of cotton cloth, of the diameter of the bottom of the picker, to which it is attached by means of a small copper wire; it should be about two feet longer than the pole. If poles of different lengths are used, then the hose may be made longer by having extra pieces to be buttoned together. To use the picker, the pole is held in one hand, while the lower end of the hose is taken in the other, and turned up to about four feet from the bottom, thus forming a loop. A fruit is detached and drops into this looped part of the hose; the next one, if allowed to fall directly against the other, would bruise both; to prevent this, the hose is grasped by the hand that holds it lower end, closing it sufficiently to break the falls, and then releasing the hose to allow the fruit to pass down. When the weight becomes inconvenient, the end of the hose is turned down, and the fruit allowed to slip gently into a basket. Mr. Ten Eyck furnished the picker complete, except the handle, to those who prefer to buy one rather than make it, but he has not patented it. (1880)

Home-Made Apple Pulper

Mr. Geo. L. Rector, Wood Co., W. Va., makes his own cider, and grinds or pulps the apples with a machine shown in figure 1. It consists of a log 12 inches in diameter, and as long as desired, with a mortise cut in the middle as shown in side view in figure 2. The upright lever or "ram" fits in this mortise and is held in place by a stout pin. There should be ¾ of an inch space between the lever and the sides of the mortise, for the

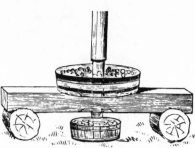

Fig. 1.—THE APPLE PULPER.

Fig. 2.—INTERIOR VIEW.

mashed apples to fall through into the tub placed below. A hopper, made of an old tub, is fitted upon the opposite of the log, and around the mortise into which the cider apples are poured. "With such a machine," Mr. R. writes, "a man can mash six bushels per hour, and work at a moderate gait." (1882)

Fruit Trees for Ornament

The question is often propounded why a farm intended exclusively for fruit should not be planted for landscape effect; why a lawn could not be planted for ornament and only fruit trees be used. The only answer is that it is eminently proper and easily practicable so to plant both farm and lawn, that these results can be obtained, and the eye, the palate, and the pocket, all be gratified.

The farmer will at once say, and truly, that to this there are two objections. The first is that he has no skill in the artistic laying out of grounds and grouping of trees, and that he can not afford to employ a landscape gardener for that purpose. The second objection is that fruit trees will not flourish when they are young without culture, that spade work is too expensive, and that he can not plow among trees irregularly grouped. To avoid, therefore, straight lines, and yet to retain regularity sufficient to allow plowing, may not be productive of the best landscape effect, but may produce an approximation to it, and will certainly give much pleasure. The accompanying diagram will show the simple method by which any farmer can plant the space around his intended dwelling with fruit trees, and yet so preserve their irregularity that from his dwelling there will appear no straight lines.

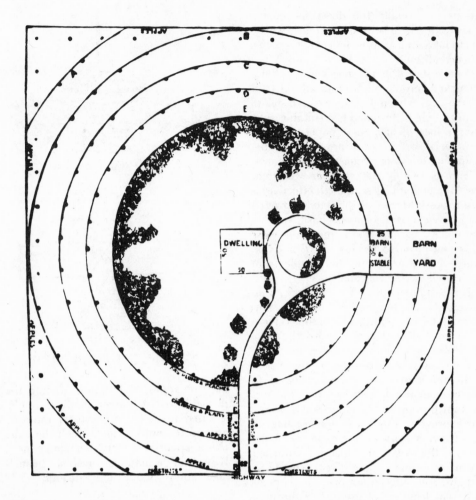

to pears on pear roots.

There are several reasons for this difference of opinion, but the chief one is the too common practice of classing the two kinds of trees and their management together, without discrimination. Dwarfs are garden trees, needing high culture, generous manuring, and judicious pruning; there are but few sorts of the pear that fully succeed on quince stocks; and the indiscriminate working of so many sorts as dwarfs, and an entire neglect of their proper cultivation, has resulted in frequent failure and disbelief in their success. Properly selected and managed, pears grown as dwarfs come soon into bearing, the fruit is easily gathered, they require little room, and will continue to flourish for many years.

The accompanying engraving is an exact portrait of a dwarf tree of the Angouleme pear, standing on the grounds of J. J. Smith, of Philadelphia, as it appeared last autumn when loaded with fruit. It was set out four years before with several others now bearing equally well, and of the same age, and was then one year from the bud. (1873)

A circle or its segment is the most pleasing line known. It is that of the sky above us, and could our eyes grasp it would be that of the earth below. The old Romans knew this when they adopted it for their arch, which the Gothic arch of the ruder ages can not equal. Let us take, for example, a plot 500 feet on each side, which will have an area of something less than six acres. From the center of each side, run a line through the plot. Take the intersection of these lines for the center of the circle. In this center plant a strong, smooth and round stake. Over this drop the loop of a rope which will extend 281 feet. To the other end of this rope fasten a strong pointed stick. With the rope extended, mark the places for the trees on the segment, *A*, and plant them before touching the interior lines. Then shorten the rope 30 feet and mark the places for the trees on the circle *B*, and plant them. Then shorten the rope 30 feet more, mark and plant on circle *C* as before. Then shorten the rope 24 feet more, mark and plant on circle *D*, as before. Then shorten the rope 20 feet more, mark, and plant on circle *E* as before. Thus the circles are all planted, and from the dwelling in the center one can scarcely see three trees in line.

The irregularity is there, and if vistas are wanted through the lines the trees can be left out which close it. Between these circles, with the planted trees to guide his eyes, a horse can draw a plow with as much ease as on a straight line. Circles *A*, *B*, and *C* can be planted with apple trees. *D* can be planted with cherries and the taller growing pears. *E* can be planted with plums, peaches, and the lower growing pears. The straight outlines of 3 sides of the plot can be planted with apples; the front side with that stately and beautiful fruit tree, the American chestnut.

Now to an eye from the house, the interior circle will have a bare, uniform, and naked appearance. Therefore this uniformity should be broken, as in the drawing, by outlying spurs and peninsulas and islands, by plantations of low growing trees and shrubs. Next to the circle could be placed a few dwarf pears or apples, or mulberries or quinces. Then could come blackberries and raspberries, and filberts and currants, and gooseberries and barberries. Then could come flat-beds of purple hazel and purple barberry, kept down by pruning. Then graded down to the level of the lawn could be the strawberries. (1880)

Dwarf Pear Trees

A controversy has arisen on the value of dwarf pears, or pears budded on the French quince stock, and kept in the form of small trees. Some persons regard them as short lived and unprofitable, and unworthy of general cultivation; while others prefer them

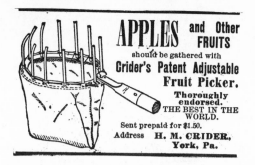
Raising Peaches

To grow them from seed, bury the pits in shallow boxes in alternate layers of sand or light earth, and expose them where they shall get a good freezing. This will crack the shell and favor their germination. Those which do not get cracked by the frost, must be broken with a hammer.

The best soil for the peach is a rich warm, gravely loam. Plow and harrow into fine condition for working. Lay off the drills 3 feet apart, and plant the seeds 10 inches asunder in the drill, and about an inch deep. This should be done as early in the spring as the weather will permit.

When the shoots appear above ground, go through the rows, and with hand and hoe give them a good weeding, otherwise they will get smothered. When about 6 inches high, put a cultivator into the rows, and then follow with the hoe. This will expedite the growth of the seedlings: in ordinary cases, the planter will have no reason to complain of the slow progress of his little trees. In many cases, budding is not resorted to; it is thought enough to grow new trees from good seed of good parent trees. But, by all means, avoid seed from trees which have the yellows or any other disease. But when it is desired to propagate any known varieties, budding must be resorted to. When this is done, the operation should be performed on the larger seedlings, in the latter part of the first season; say, from the first to the middle of September. As the peach grows quite late into the Fall, this will give the bud time to unite well with the stock before winter. Next spring, before growth begins, head back the stock; the bud will soon start, and grow several feet in the season. It will be ready to transplant in the Fall.

A word in reference to the situation of the peach orchard. To guard against late spring frosts, avoid low, rich ground, and choose rather high and rolling land. Here the trees may grow less luxuriantly, and the fruit ripen a little later, but frost and various diseases will make less trouble. It is a mistake, however, to suppose that the peach is partial to a poor and thin soil. It is often put into this, as in Delaware and New Jersey, because this land will yield no other crop so remunerative. It is chiefly because of the poverty of the land that the peach orchards

there are so short-lived, often flourishing no longer than 3 or 4 years. Give the peach a good, substantial loamy soil, well drained and moderately enriched, and it will very often remain productive 15 or 20 years.

In setting out a young orchard, choose fresh land. Plant the trees 16 feet apart unless you know that they are likely to live more than 12 years. Give the land a light annual dressing of manure, and keep it well stirred. After the trees have begun to bear plentifully, no other crop should be taken from the land.

They may be pruned during the growing season to advantage, by pinching in the shoots in such a way as to give a compact and uniform top, and keep the fruit more within reach. The tendency of the sap, the most rapid growth, and much of the fruit, are always towards the ends of the limbs, and thus, each year, our pruned trees stretch away further and further from the ground. Pinching in gives larger and better developed leaves, better buds for next year, and greater perfection to the fruit, which is borne both on year old, and still older wood. The blooming is so profuse that little fruit is lost by pruning. When thus shortened in spring or autumn, better the former, cut back, leaving only four or five good buds on wood of the past season's growth, and remove at the same time all side shoots that fill up the tree too much. (1862)

COMPARATIVE FORMS OF PEACHES.

NUTMEG.   EARLY ANNE.   EARLY TILLOTSON.

SERRATE EARLY YORK.   LARGE EARLY YORK.

JAQUES' RARERIPE.   EARLY CRAWFORD.

Preparing Peach Stones

When the stones are kept dry all winter and then planted in spring, very few, if any, will germinate. In nurseries, the stones are not allowed to get very dry, but they are stratified or bedded just before winter sets in. The usual manner is, to mark out the limits of the bed and spread the stones over it to the depth of two or three inches; the stones are then spaded in as if turning under a dressing of manure. By this operation they are distributed through and well mixed with the soil, where they are left to freeze and thaw all winter. This treatment causes the halves of the stone to separate and the pressure within of the swelling seed can push them apart.

This may be effected by other methods than that of spading in. The stones are sometimes spread upon the ground and covered with spent tan-bark or sawdust to the depth of three to five inches, and thus exposed to the weather. In spring, when the ground is ready to plant the stones that have been spaded in are separated by throwing the soil of the bed upon a riddle, such as is used by masons; the earth falls through while the stones are left upon the riddle. Those that have been under tan or sawdust are more easily recovered at planting time.

It is to be assumed that those who at this time ask what is to be done "to prepare peach stones for planting," have as yet done nothing with them, and that they are still dry. Such should at once be mixed with sand, or sandy earth, in a box, and placed where they will be exposed to all the changes of the weather. This will place them in a condition similar to those that have been spaded in, but have been dried for some months it is likely that a share of them will remain unaltered, and that in spring the halves will remain firmly together. Such stones must be carefully cracked, holding them between the thumb and finger upon a block and striking the edge with a hammer; the kernels being thus removed, are to be mixed with earth or damp moss, and kept in a warm place until they germinate. The stones that have been bedded or otherwise exposed to the action of frost, and are still unchanged, are not planted with the other, but are separated from those that have begun to germinate, and cracked before they are planted. (1882)

Cider—Stimulants—Cider Vinegar

Cider making in the olden time was reckoned one of the important parts of farm labor. A goodly row of cider barrels must be ranged beside those of beef and pork, and the majority of farmers would almost as readily have dispensed with one as the other. The cider pitcher stood regularly upon the table at meal times, and the jug was a constant field companion. Happily that day is past.

Experience has proved that more and better work can be done on the farm without than with alcoholic stimulants. (Fermented cider always contains a large amount of pure alcohol.) Every exaltation of the feelings is followed by a corresponding depression, so that in the end there is no gain of strength from stimulants, while the system is more rapidly worn out by these changes from the normal state. In surgical operations, or in cases of extreme sudden depression, it is sometimes necessary to borrow a little from the future, by resorting to temporary stimulants; but as a rule no permanent good can come from them to a person even in but moderate health.

Good nourishing food—like beef steak cooked rare— well masticated, adds positive strength, and is the best of all stimulants. But supposing cider to be uninjurious for the time being, there is always a great danger that the use of this, or of any of the beers, which are flat and stale unless they contain more or less alcohol, will develop a taste and craving for stronger alcoholic drinks.

The general belief of the above facts has so reduced the amount of cider manufactured, that for years past not enough has been made to furnish a supply of vinegar. It is difficult to obtain a good quality of the pure article in most of our cities. The market is filled with unwholesome compounds, made from the refuse of distilleries, and with acids of various kinds. None of these deserve the name of vinegar, while in some cases they are absolutely poisonous, or at least ruinous to the teeth. Those who have continued to make cider for vinegar, find a ready sale at good prices, and probably apples may in this way be turned to better account than by feeding them to stock, although they are by no means unprofitable when fed.

The requisites for making good cider vinegar are few. Sound apples free from dirt, ground fine will yield the staple. It is well to arrange for two sorts, "extra" and good; the former to be made from the first pressings of the pomace, by which is obtained all the juice possible to be extracted without watering. After this pomace is removed from the press, pour a few pailfuls of water upon it, and shovel it over occasionally for a few days; then press it again, and add two or three quarts of inferior molasses to a barrel of the juice. Keep in a warm situation, and it will ferment into fair vinegar—far better than much of the trash sold under that name.

Vinegar is usually produced by simply allowing cider to stand in barrels with the bung hole open, until the second or acetous fermentation is completed, and it requires several months for the process to be finished, the time varying considerably with the degree of temperature. As fermentation depends upon the absorption of oxygen from the air, it will be much hastened by exposing as large a surface as possible. In some establishments where the manufacture is largely carried on, this is accomplished by allowing the liquid to drip slowly through beech wood shavings; the work of oxydation is finished within a few days instead of extending over months. The old-fashioned plan, however, will answer all purposes of ordinary vinegar making. (1862)

Making the Best Cider

To make good cider, the apples for each pressing should be nearly as possible of one kind, free from rot, leaves, and all foreign substances, that the fermentation may be complete and uniform. Apples should be selected, the juice of which has the greatest specific gravity, as such juice contains the most sugar, and makes the richest cider. They should be ground and pressed with scrupulous cleanliness, and every step of the process, from the gathering of the fruit to the final barreling and bottling the liquor, should be conducted in the same careful and unexceptional manner.

The apples should ripen upon the tree, be gathered when dry, spread in an airy covered situation for a time, to induce any evaporation of acqueous matter, which will increase the strength and flavor of the liquor; and finally they should be separated from rotten fruit and every kind of filth before they are ground.

The apples should be reduced by a mill to a uniform mass; to give it color, the pomace may be exposed to the air from twelve to twenty-four hours till it becomes red; then press out the juice slowly; put it in casks; bung it up and immediately place it in the cellar, leaving out the bung. Fill up the cask to the bung, in order to let the impurities flow over. Before the fermentation ceases, insert a flexible tube through the bung, block tin will answer, and bend the other end, like a syphon, into a cup of cider, or water placed on the cask near the bung, to allow bubbles to escape. (1870)

BALDWIN

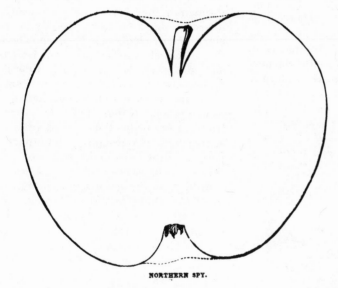

NORTHERN SPY.

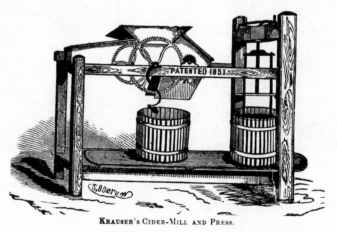

KRAUSER'S CIDER-MILL AND PRESS.

Apples differ not only in their flavor, color, and time of ripening, but in the proportions of their constituent parts. The characteristics of a good cider apple are: a red skin, yellow, and often tough and fibrous pulp, astringency, dryness and ripeness at the cider making season. When the rind or pulp is green, the cider will be thin, weak and colorless; and when these are deeply tinged with yellow, it will almost always possess color, with neither strength nor richness. The apple, like the grape, must attain perfect maturity before its juices develop all its excellence; and, as many of our best eating apples do not acquire this maturity until winter or spring, such fruit seldom yields good cider. In a dry apple, the essential elements in cider are generally more concentrated, or are accompanied with a less proportion of water, than in a juicy one; of course, the liquor of the former is stronger than that of the latter. Of our best cider apples, from ten to twelve bushels of fruit are required for a barrel of cider, while of the ordinary juicy kinds, eight bushels generally suffice.

About Quinces

Here is one of the old-fashioned fruits which deserves all that has ever been said in its praise. It is known botanically as *Cydonia*, from the city Cydon, in Crete, where it seems to have first attracted attention. As a stock on which to graft the apple and pear, it has become quite serviceable, giving the fruit great precocity, increased size and improved flavor. Then, who does not know by experience that its own fruit when stewed, or mixed with apples in pies and tarts, makes a palatable dish? Says an old gardening author: "When apples are flat, and have lost their flavor, a quince or two in a pie or pudding adds a quickness to them." To make a first-rate apple pie, use one fourth quinces sliced with three fourths apples, and add a few thin slices of candied lemon peel or citron; and more than a simple allusion to the delicious preserves, jellies, and marmalades made from the fruit in its purity, is simply an aggravation.

From the kitchen, let us now go to the garden. The quince may be propagated by cuttings or layers. The cuttings are prepared in the same way as for grapes, they should be set out in a deep, light, well-worked soil; if a little shaded, success will be the more certain. Let the rows be 18 inches apart, and the plants 6 or 8 inches apart in the row. Keep the ground well stirred during the summer, and mulched in very dry weather. Layering is much practiced, and is a surer method than by cuttings. Shoots are bent down, slightly cut, and covered with earth. Another method is to earth up the mother plants in the spring: shoots will spring up from the old stocks, with roots attached.

In setting out the young plants in their final position, place them 10 feet apart, and give them the best culture. As the quince is often found wild by the side of streams, it has been supposed by some that it needs a damp soil, and is often set in the lowest and poorest part of the garden. A mistaken practice. Give it an open situation, a deep soil, and let the ground be surface manured every other year.

The fruit will be abundant and large and fair, instead of scanty and knotty, as seen in its wild condition. Bushes well planted, and generously treated, will begin to bear fruit the third year, and will yield good crops for 30 years. First class specimens will bring two cents a piece in the New York market. The only pruning required is to cut out the old and twisted or decaying branches. In addition to the manure, given an annual dressing of salt, just enough to white the surface.

There are only two varieties extensively cultivated: the apple or orange quince, and the pear-shaped. The first is the most popular, as it bears abundant, fair, orange colored fruit. The second is dryer and firmer than the other. It does not become as tender when cooked as the other, but ripens a fortnight later, and keeps longer into the winter. (1862)

Espalier or Wall Training of Fruits

Whoever has not practiced the close pruning and training of fruit trees upon trellises or walls shrinks from attempting it, as from a great labor likely to be unremunerative. But whoever begins, even in a very humble degree, becomes almost uniformly an advocate of the practice.

The currant is probably the easiest of all fruits to experiment upon, and from practice with it the general principles of close pruning may be learned, while the results are most gratifying. Last year George H. Hite, of Morrisania, N.Y., placed upon the exhibition tables of the *Agriculturist*, some extraordinary stalks of currants, which we condense an account of his process. He fruits his currants on single canes, on spurs, which he causes to form of nearly

uniform strength from the ground to a height of several feet; and thus these long canes are perfectly clothed with large fine fruit.

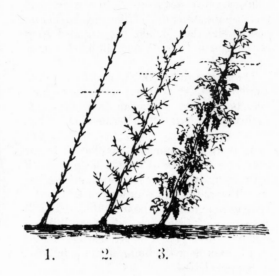

1. 2. 3.

In fig. 1 is shown a strong shoot thrown up from the ground at the root of an old bush, or an independent cane. At the end of the first season, it is cut back one third. If attached to a trellis or otherwise made to incline the next spring, the buds will all break uniformly from top to bottom. If it stands upright, the sap will flow chiefly to the uppermost buds. When the buds have grown a few inches, as shown at fig. 2, they are pinched off to leave spurs about an inch long, as shown enlarged in fig. 4. These will develop a rosette of buds

Fig. 4.

at the base of each, which are the fruit buds for the next season. The terminal bud at the top of the cane is left, and will have stretched upward considerably. In the autumn shorten it in one fourth of its growth; and the succeeding years still less. The next year the cane will bear full, as shown at 3, and if the soil be good, and the bush has air and light, the fruit will be large and fine. The spurs must be prevented from growing long, and they will bear year after year. Several such canes may be grown from an old root, while the old bush, by stopping in the spurs, will bear much better until the new canes are ready for full bearing. Currants may be set

out about a foot apart, and thus in small compass yield an immense quantity of fruit.

Apples and pears may be subjected to very similar treatment, being planted close, or further apart, according to the system followed or the height to which they are wanted to grow. Fig. 5 shows an experiment conducted

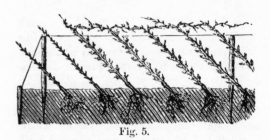

Fig. 5.

by Mr. F. Otto, in our own garden, which promises success. Dwarf apples are planted 1 foot apart, and trained to a rod 18 inches above ground. (A wire is perhaps preferable). The inclined position causes them to break throughout their length quite uniformly, and to secure this a greater or less degree of inclination or curving may be given. The spurs if kept closely pinched in, form fruit buds. The tops when they have reached the height desired are allowed to branch somewhat, and the branches are inarched or grafted together by approach, so that after a few years the fruit-bearing hedge thus formed will be self-supporting and substantial and ornamental. Keeping fruit trees under this close control enables us to secure healthy, well developed fruit buds on the spurs, and to protect easily the tender kinds from frosts. Peaches, plums, and berries do very well upon walls or trellises. (1862)

The Bayberry or Wax Myrtle

Near the coast of the sea, and of our great lakes, is found growing in almost every variety of soil and situation, a low and rather irregular shrub, known as the bayberry. It is quite dwarf and stunted in all its parts, when it grows in the sandy soil of the shore, but when it is found along the borders of marshes, it is much more luxuriant. The usual size of the leaves, and the general aspect of one of the smaller branches are of a fine, shining dark green, and thickly sprinkled over with minute resinous dots.

When slightly rubbed, the leaves give off a pleasant balsamic fragrance. The staminate and pistillate flowers are borne on different plants, both kinds are produced in small cone-like scaly aments or catkins, and not at all showy. The fertile flower clusters produce several small one-seeded berries, or more properly nuts, which are at first green, but at maturity they are covered by a whitish granular powder, which is wax.

This shrub extends from Nova Scotia to the Gulf of Mexico, and in some localities is turned to considerable profit. Its fine green leaves, which do not readily wither, are extensively used in making up the bouquets sold in our city streets, and are one of the most available greens for this purpose. The great value of the plant, however, is in the wax with which the berries are encrusted. The berries are boiled in water and the wax melts, rises upon the surface, and may be dipped off or allowed to harden there as the water cools.

Where the shrub abounds, the wax, or "bayberry tallow," as it is frequently called, is collected in considerable quantities for domestic use and for sale. The wax is greenish white, has a slight odor, and is more brittle, and has a more greasy feel than beeswax, and it melts at a lower temperature than that does. It is used for making candles, either alone or mixed with tallow. When mixed with tallow it gives greater firmness, and the candles in burning diffuse a pleasanter odor. The wax is used in some preparations for leather, and it is the material employed for stiffening the ends of circular lamp wicks. Another species, *Myrica Gale*, the Sweet Gale, found in wet places, has less fragrant foliage, and its fruit does not furnish wax. (1866)

Blackberries & Raspberries

Everyone who has gathered wild blackberries is aware that the stems grow to the height of six or eight feet, and gracefully bend over at the top. The lower part of the cane bears little or no fruit, it being nearly all at the top of the bush. The same happens on a smaller scale with the raspberry when left to itself. While we cultivate the blackberry for the sake of better fruit than the wild plants usually afford, we should also endeavor to have more of it, and more conveniently placed on the bushes.

From questions that are asked, it is evident that all are not aware that the stems of the blackberry and raspberry (at least those cultivated for fruit) are only biennial. The plant throws up from the root, often at some distance from the old stems, vigorous shoots which grow rapidly, and by autumn will become ripe and hard canes, like the old ones. The old canes, which have given a crop of fruit, have completed their work, and though they may remain alive for awhile, will all be dead by next spring. When the fruit has been gathered, it is best to cut the canes entirely away, to give room to new ones. These should have more attention than they usually receive; if left to themselves, they will become just like the wild plants, straggling, and with their fruit all at the top. Not only on account of the greater quantity of fruit, but

LAWTON BLACKBERRY.
For the original variety, for Circulars free, address
WM LAWTON, New Rochelle, N. Y.

for the ease in picking it, should the canes be pruned. Blackberry canes should never grow over five feet, and many prefer to keep them at three feet. Whenever the green shoot has reached the desired height, remove the top, or growing point, which, being tender, may be pinched off with the thumb and finger. Soon after this is done, branches will start along the stem, and these should also be pinched, the lower ones when 18 inches long, and the upper when 12 inches. By a little attention, once a week, or oftener, giving the needed pinching, the blackberry, instead of being a long straggling shrub, catching at the clothing of all who approach it, may be brought into the form of a neat pyramidal bush, which, the next season, will be loaded with fruit from bottom to top.

The same treatment may be followed with raspberries, which are usually kept shorter. One of the greatest pleasures in gardening is found in training and shaping plants, and making them grow as we wish, and in nothing are the effects of this more strikingly shown than in the blackberry and raspberry. (1882)

Cranberry Culture

"M.H.P.," Augusta, Me., asks if muck is essential to successful cranberry culture, and if ordinary "black soil" will not answer as well. This question comes to us from several readers each spring. While we have no doubt that one, if sufficient pains be taken, can raise cranberries in a garden bed, upon ordinary soil, the "sufficient pains" would be utterly impracticable upon a profitable or commercial scale. While we have known small quantities of the fruit to be raised upon upland, we have no knowledge of successful, continuous culture anywhere outside of a regular cranberry bog.

In the first place, there must be a deposit of muck or peaty soil; in some localities this muck is not more than two feet deep, underlaid by sand, which may be brought to the surface by a system of trenching; but where the muck is six or eight feet deep, sand must be brought from elsewhere, to cover the muck, after the native growth has been removed, to the depth of four to six inches. Muck and sand are essentials, but these must be so situated that they can be drained of standing water for 12 or 18 inches below the surface, by means of ditches, and more than this, there must be a supply of water at command, by which the whole surface of the cranberry field can be flooded, at once, to completely cover the plants with water, and from which the water can be drawn off as suddenly.

It will be seen that there are not many localities which will meet all these requirements. Successful cranberry growing depends upon the selection of these, and the failures have come through attempting to grow the crop where some one of the conditions was lacking. There are several localities at Cape Cod, in Massachusetts; in several counties in New Jersey, in parts of Minnesota, and in other States, which seem especially adapted to cranberry culture.

The reason for sanding does not seem to be generally understood. When the natural growth is cleared from the surface of a muck bed, the few plants that spring up and interfere with the growth of the cranberry vines will be from the seeds that may be already in the muck. When the surface is covered with sand, and the cranberry plants are set in it, they are in a condition most favorable to their growth; the sand at the same time keeps down all other plants, and the cranberry has a chance to spread and occupy the whole surface, forming a mat which will keep down all other growth. Natural cranberry bogs are often brought into abundant bearing by simply giving a heavy dressing of sand. (1882)

How to Grow Gooseberries

Many cultivators suffer from insects and mildew so badly, they have about given up the attempt to raise this very agreeable fruit. We suspect that a barren soil, stunting the growth of the plants, is, in many cases, the cause of the blight complained of. Another cause is the sudden alterations of temperature that occur almost every summer. It is a mistaken notion that because the gooseberry is often found wild in poor soils, it therefore needs no manure.

The treatment which ensures the best results is as follows: Give the plants a dressing of manure in the fall, packing it around the roots in spring. Keep the ground clean and open until about the middle of May or first of June. Then, spread under the branches a layer of straw 5 or 6 inches thick, letting it extend over the ground as far as the roots penetrate. This mulching should remain on the ground until the first of September, when it should be removed and the soil worked clean. The design of this midsummer dressing is to prevent any check in the growth of wood or fruit, and to keep the air about the bushes uniformly moist and cool. In this simple way, we manage to get good crops, as often as five years out of seven. Persons near the seaside might use sea weed or salt hay for a mulch. Tanners' bark is often used with success. (1862)

Cultivation of the Huckleberry [Blueberry]

When Bartholomew Gosnold, in 1602, discovered wild grapes growing in great abundance in the swamps and low grounds on a little islet near the New England coast, he gave to it the name of Martha's Vineyard, no doubt believing that he had found the home of the wild grapes of the New World. But neither that little island, now known as "No-man's Land," nor the larger island, which bears the name of Martha's Vineyard, are considered favorable locations for vineyards, although the wild grapes do grow all along the New England coast, and in swamps and low grounds throughout these United States. While it is true that the wild grapes of North America are found more abundantly in swamps and low grounds, than on high and dry soils, still, no vineyardist would think of planting a vineyard in a swamp, because long experience has shown that high, dry and well-drained soils are far preferable for such purposes than those that are low and wet.

There is another very valuable native fruit, about which the same erroneous ideas exist, that were for a long time held in regard to the indigenous grapes; it is our swamp high-bush huckleberry, or blueberry (*Vaccinium corymbosum*). It is found growing wild in the same localities, and under the same conditions as the wild grape, not only in swamps, but also on high and dry soils. Because the plants are more abundant in swamps does not prove that under cultivation, low, wet soils would be the best. From my own experience with this species of the huckleberry, I should not choose low, wet soils in which to plant it for fruit, but in a sandy, or, at least, well drained one. The plants thrive best in peat, and the almost pure vegetable deposits of the swamps; also in the light, sandy soils, and even high up in the hills of New Jersey and adjoining states, in light, sandy soils, in which the running blackberries and five-finger plant, have to struggle to obtain nutriment from the sterile soil. A plant that will grow and thrive—bearing a crop of fruit in moderately favorable seasons—in such soils, will certainly thrive under good cultivation, provided the soil is not a heavy, unctious clay. I have had no experience in cultivating the huckleberry on clay soils; but in sand, or sandy loam, they may be grown almost as readily as currants or gooseberries.

The plants can be had in abundance from the open fields or swamps, and usually they can be lifted with good roots, and then by cutting away the older stems—leaving the younger and more thrifty—there is no difficulty whatever in making them live. They may all be propagated by layers, or seed; but the latter is a slow process, as the plants make little progress for the first few years, and we may save a decade or two by taking up the wild plants.

As there are several distinct natural varieties of the high-bush species, as well as of other species, it is as well to mark the plants to be taken up when in leaf or fruit. The genuine, or true *Vaccinium corymbosum*, bears quite large, round berries, covered with a blue bloom; but there is a variety with oval fruit, jet-black, without bloom, and another with globular berries and also destitute of bloom. Of the dwarf, early blueberry (*V. Pennsylvanicum*), common to high, dry and rather sterile soils, there are also several distinct natural varieties; one of which is an albino, the fruit being pure white, and fully as transparent as the white grape currant.

In cultivating any of the huckleberries on sandy soils, it is advantageous to keep them well mulched, thereby insuring an abundance of moisture at the roots, as well as preventing any baking, and overheating of the surface soil. Under proper care and in rich soils, the plants will grow far more rapidly, and yield larger crops of fruit, than when left to grow uncared for, as in their native habitats. (1886)

WALKER'S SEEDLING.

Mulching the Strawberry Bed

A mulch is placed on the strawberry bed for several purposes. It is not needed in the fall until freezing weather has set in, its first use being that of keeping the ground from thawing during the warmer days of winter and early spring. Frequent freezing and thawing heaves and cracks the ground, breaks the roots of the plants, injures the crowns, and often kills the plants outright. The mulch must furnish a coat of sufficient warmth, to prevent such damage as far as possible. At the same time, it must not be spread on too thickly, as it will then kill the plants by smothering them. A coarse mulch may be spread more thickly than a fine one, hence is better for winter protection.

The next use of the mulch is to prevent the starting of the plants too early in the spring. It may be desirable to retard the growth of the vines for two purposes. In the first place, it is not safe to let the plants blossom very early, on account of frosts which may occur at this time. And, secondly, it is often desirable to hold back a portion of the crop as late as possible, in order to lengthen the bearing season. A coarse mulch admits the air to the plants more freely, and can be kept on later in the spring than a fine one, and is, therefore, better for this purpose, also.

IOWA.

The next, or summer use of the mulch, is to fill the spaces between the rows, where it serves the triple purpose of keeping the earth soft and moist, smothering the weeds, and keeping the berries clean. Here is a need of something that will stay where it is placed. A mulch that has kept its place all winter may become quite unmanageable when the ice has released its tenacious grasp and the material has become thoroughly dry. Especially, when stirred up and drawn into central ridges where wanted, is some strong wind likely to play mischievous pranks with it. I have had a few sad experiences along this line, in using straw for a mulch. One spring, after a dry spell of some weeks, a violent wind removed the entire mulch from a large bed and scattered it over the garden, field and lawn, for a distance of forty rods or more. Sometimes a strong wind comes in the fall, soon after the bed has been covered, and the mulch must cling to the ground quite strongly or it will be carried off. Having tried straw, leaves, coarse manure, corn fodder, and various other materials, I am satisfied that coarse marsh hay makes the best mulch for the strawberry bed. This serves the various purposes admirably, and it has no seed that will prove troublesome. On many swampy or low places may be found coarse kinds of grasses unfit for hay, which may furnish the strawberries with just the covering they need, and may be obtained without cost save that of cutting and getting in place. It takes from two to three tons of dry material to cover each acre. One can economize by placing the mulch along the matted rows, and leaving intervening spaces bare; but it is better to cover the ground. Coarse hay when wet dries more quickly than straw, and hence does not decay so quickly. The keeping of the dirt and sand from coating the strawberries will alone pay for mulching, but the weeds are killed and the soil kept moist also. (1893)

If these new plants are left where they grew, the oldest of them will bear fruit the next season. But if we can manage to remove these new plants without disturbing their roots and checking their growth, the oldest of them will produce an even better crop. This can be done by what is called "pot-layering."

The method is very simple, anyone can readily increase his strawberry beds in this manner. Small pots, not over three inches across, are filled with good soil and plunged in the strawberry bed, just under the bud, at the end of a runner, so that its roots, instead of striking in the soil of the bed, will push into the soil in the pot. The pot should be set down in the bed, so that its edge is not above the general surface, else it may get too dry. As the wind may blow the runners about, it is well to lay a small clod on them, to hold them in place, or a small hooked twig may be used. In two or three weeks the new plant will be sufficiently provided with roots to allow the runner that connects it with the parent plant to be severed. It is then to be taken to the new bed, the ball of earth containing the roots turned out and placed in a hole made to receive it. It will go on and grow without any check.

To prevent any injury from drouth, it will be well to mulch the plants with a little hay or straw as they are set out. The sooner the new plants can be transferred to their new bed the better; if it can be done in August, a full crop may be expected next year. (1882)

POT-LAYERING STRAWBERRIES.

More Strawberries

Those who wish to increase their strawberry plantation, should set about it at once. The strawberry plant is a most generous one; besides giving a crop of fruit, it also affords a crop of plants. Long slender stems push out from the crown of the plant, and a bud forms at the end; roots soon appear to fix this bud to the soil, and it becomes a new plant; another runner pushes from this, and so on. A strong plant will, in a single season, surround itself with a colony of young plants.

Historical Notice of the Isabella Grapevine

The Isabella grapevine, so celebrated throughout the United States, for its hardiness, vigor of growth, and abundant yield of fruit, it is highly probable, is a hybrid produced by cross fecundation between the vine of Europe and one of our native species. Concerning its origin and history, I am indebted principally to General Joseph Swift, U.S.A., of Geneva, New York, for the following account, which I trust will be no

less acceptable in coming from so respectable a source, than in the interest elicited in so valuable a production.

It appears that General Smith, of Smithville, North Carolina, in 1808, procured from Dorchester, South Carolina, several roots and cuttings of a hybrid vine, which, it is said, had been originated there by some families of Huguenots, between the Burgundy grapevine from France, and the native fox grape *(vitis labrusca)* of that vicinity. In the year 1817, a vine produced from these cuttings, was transplanted from Smithville, by Mrs. Isabella Gibbs, in honor of whom this variety was named, to the garden then owned by her husband, Colonel George Gibbs, which was situated along the southerly side of Cranberry, between Willow and Columbia streets, in Brooklyn, New York. In 1819, the garden was purchased by General Swift, who very generously distributed roots and cuttings of this vine among his neighbors and others, more especially to the late William Prince, of Flushing, Long Island, by whose efforts it became widely disseminated throughout the Union, and was sent to several countries in Europe, Madeira, &c. The garden has since been divided into lots, and partially occupied by buildings, and the original Isabella vine, after attaining a circumference of more than a foot, was severed to the ground, about the year 1837, in order to make room for the improvements going on at that time. Portions of the parent stock, however, are still growing in perfect perfection, and annually produce an abundance of fruit. Mr. A. G. Thompson, their present owner, informs me that, in grading the lots on which they stand, it became necessary to raise the surface some two or three feet, and that the original roots are still supposed to remain at that depth in the earth, a conclusive proof of the advantages derived from deep planting in a free and open soil. (1845)

Layering the Grape Vine

While many varieties of the grape will grow readily from cuttings planted in the open ground, others need to be placed in a propagating house, where they can have bottom heat, and a few varieties can hardly be grown from cuttings at all, but must be layered. Layering consists in burying a portion of a cane in the soil, in order that it may strike root while it is yet attached to the parent vine; when a sufficient number of roots have formed, it is cut away from the old vine and set out as an independent plant. This method is of necessity employed by the nurserymen for those varieties that will not readily grow from cuttings, and it offers to the amateur the most certain means for the propagation of all vines, as he does not care to multiply them in large numbers.

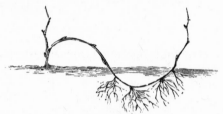

A GRAPE-VINE LAYERED.

The best method of layering is to use in early spring a cane that grew and ripened the previous season. If it is now too late for that, one can layer the shoots of the present season's growth. The shoots for this purpose should be those which start near the base of the vine, which have been allowed to grow for the purpose. They will usually be hard enough by mid-summer, for if layered when too soft and green, they might decay. Having ascertained, by extending the shoot, the best place for burying a portion of it, dig a hole with the spade, and carefully bend the shoot, in order that a part of it may be buried to the depth of six inches or more, cutting away all the leaves from the buried portion. Press the earth firmly down over the stem, and tie the portion which extends above ground to a stake. It is well to put a mulch of some kind over the layered portion; there is nothing more effective to retain the moisture than a few flat stones. Usually abundant roots will be formed by autumn, when the new plant may be severed, or it may be left until early the following spring. (1882)

The Grape Vine in Summer

Those who grow grapes for profit, or have a large number of vines, will of course have fixed upon some method of training them. and understand how this is to be carried ou.. All such will do well to take some standard work on the vine as a guide. Where one has but a few vines, and these often grown as much for the shade as for the fruit, he does not usually care to make a special study of the matter; still, by attending to a few simple rules, he can with a very little trouble greatly increase the quantity as well as the quality of the fruit, and at the same time keep his vines in a satisfactory condition.

In June the time for pruning the old wood is past, and the vine now presents a greater or less number of canes upon which at intervals are green and tender shoots, that have started in the spring. Of course, we now speak of the vine in general, no matter how it has been trained. As one stands before the vine, he must consider that every one of these young shoots now just growing will, if left to itself, make by autumn, just such a ripened cane, as that from which it starts. Will there be too many of these canes? Will there be a great crowding and tangling if they all grow? If so, break a portion of them off while they are still young. If a bud produces two shoots, and they are often double, always break out the weaker one. Each shoot on a vine old enough to bear, will appear as in fig. 1. Beginning

Fig. 1.—SHOOT ON OLD VINE.

below these is a leaf, with opposite to it a cluster of buds; a short distance above, another leaf with another cluster of buds, but their positions reversed, the second leaf being over the first cluster, and so on, alternately, some vines producing three and some four clusters of buds, and beyond these, instead of buds, there will be merely a tendril opposite the leaf. Observe that the whole growing vine is only a repetition of this; a portion of stem, a leaf with opposite a cluster or a tendril, another length of stem, a leaf and a tendril, or cluster, and so on to the very end.

That point of the shoot, where the leaf and opposite tendril or cluster are attached, is called the node. A single node is shown at fig. 2. When it is understood that every vine, however confused it may appear, is made up merely of repetitions of that which is shown in fig. 2, the matter becomes much simplified. The tendril may be the stem of the fruit cluster, or it may be bare and used to hold up the vine. When it first appears, the tendril is nearly straight, except at its slightly hooked end, when that end comes in contact with, and catches hold of some object, it then coils in a manner most interesting to witness. When it bears fruit, there is often, as in fig. 2, a bare branch ready to catch hold of something, and help sustain the weight of the cluster.

It will be observed that in the axil, the place where the leaf-stalk joins the shoot, two buds will soon appear, which we will notice presently. If the shoot in fig. 1 is left to itself it will continue to grow and often reach several feet beyond the uppermost cluster. With every vine, no matter how it may be trained, it is a safe rule to stop the growth after one or two leaves have been formed above the uppermost cluster. If any one is afraid to do this, let him try it on a few shoots and leave the rest to grow. This should be done while the shoot is still tender, and its end can be pinched by the use of the thumb and finger nails. The leaves remaining upon the shoot will soon become much larger and thicker, while the fruit will also grow much larger than if left unpinched. This is a general rule to be followed on all bearing vines. In fig. 1 a line is given to show where to cut in case the vine is disposed to bear too much, a point we need not consider now.

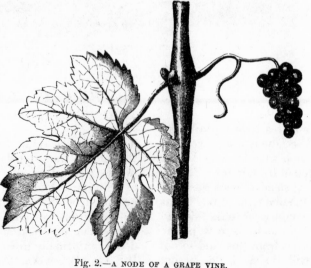

Fig. 2.—A NODE OF A GRAPE VINE.

Fig. 3.—SHOOT WITH LATERALS.

Soon after the top of the shoot has been pinched, side shoots will appear, as shown in fig. 3. These side shoots are for convenience called laterals, and one will appear from the axil of each leaf. If these are allowed to grow the vine will soon become a tangled mess; but they must not be broken off, altogether, but should be checked by pinching. As shown in fig. 2, there are two buds in the axil of the leaf; the lateral comes from one of these, and if this lateral be broken out, the other bud will start into growth and produce another one; this is to be avoided, as that bud may be needed next year; consequently when the lateral has produced two or three leaves, it is pinched off, leaving only its lower leaf; if so, that is to be pinched again, in the same manner, back to its first leaf. This is made plain by fig. 4, which shows a lateral that has been

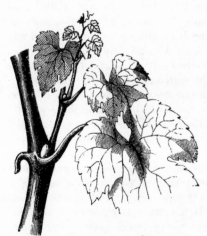

Fig. 4.—MANNER OF PINCHING.

pinched at *a*, and a second growth has started which should be pinched back to its lowest leaf at *b*. Although it takes some space to describe these operations, they are really very simple, and if one will observe the directions the fruit will be greatly improved, and vine prevented from becoming a confused mass of stems and foliage. Of course all the insects observed should be destroyed. (1882)

The Grapevine in November

Take one cane by itself, whether there are two or three upon a young vine, or a multitude upon an old one, examine and understand it, and see what it will do next year, and its treatment will become plain. Suppose we did not prune, what would happen? If the cane, fig. 1, were left as the fall of the leaf

Fig. 1.—YOUNG GRAPE SHOOT.

found it, its buds would next spring push into shoots. The uppermost buds, according to a well known law of growth, would start first, and push their shoots, the next lower ones, starting later, would produce smaller shoots, and there would be a series of smaller shoots as we go down the cane, as in fig. 2, while the lowermost buds, being deprived of sap by the upper shoots, may not start at all.

It will be seen that the most vigorous shoots will be from the upper buds, while the lower buds, which were formed earliest in the season, and are most likely to contain the rudiments of fruit, are quite starved and useless. Each year, if the vine goes unpruned, the new growth will be further from the root, and the best developed buds, starved, as it were, by the more rapid growth of those

Fig. 2.—BADLY PRUNED BRANCH.

above them, and but very little fruit will be produced. Suppose instead of allowing the cane, fig. 1, to go unpruned, we at this season cut it off as shown by the line, leaving but two buds. Next spring, these two, having all the nourishment that would otherwise have gone to the buds above them, will push with great vigor, and being the oldest and best developed, will probably bear fruit abundantly, and next autumn would present the appearance of fig. 3, as contrasted with that of fig. 2. We prune then, to be sure of fruit, and to avoid growing a great number of shoots for which we have no use. Now if one understands this as to a single cane, the pruning of the whole vine is plain.

Fig. 3. WELL PRUNED BRANCH.

If there are too many canes, so many that even one or two shoots from each will cause crowding, let him cut them out altogether. Then cut the canes that are allowed to remain, back to two buds each, keeping in mind the fact that each of these two buds will produce a shoot, which next autumn will appear as a cane, just like the one he has now before him. In pruning it is rarely desirable to provide for more than two shoots from a cane, but it may be well, especially in severe climates, and to guard against accidents, to leave three or four buds in pruning at this season, and then late in February, or early in March, at any rate before the sap starts, cut away the extra buds. (1882)

A Convenient Grape Trellis

For some three years past, I have used the trellis described below, and have found it to be just the thing. When the time comes to cover the vines, I remove the support of the trellis, and lay it flat on the ground. The vines do not have to be disturbed at all. In

A HINGED GRAPE TRELLIS.

spring, when the covering is removed, the trellis is raised, and the supports put in place. There is no tying up of vines or branches with this trellis, consequently a saving of time, labor, and vexation. This trellis can be made to stand perpendicular or slanting by changing the position of the posts supporting it.

Take two pieces of plank, and bore holes through the upper part of them, to receive the bolts which are to hold the trellis in place. Drive these pieces into the ground, as far apart as the trellis is long. The trellis can be made of any convenient size, and consists simply of five stout pieces of wood, nailed firmly together with long cross-strips, to furnish support for the vines. At the bottom of the end pieces, bore two holes to correspond with those in the pieces of plank, and put an iron bolt, fastening with a nut, through these end pieces and the planks, thus making a hinge upon which the trellis can be made to move easily. The trellis is so simple that any man can make one in a short time. (1882)

A Circular Grape Trellis

Those who understand the laws governing the growth of the vine can train it in a great variety of ways. A subscriber in Manchester, N.H., writes as follows: "I send you a sketch of a trellis that I have used in my garden with satisfactory results, both as an ornament and support for the vine. It is not patented, and one can make it who chooses, as follows:

Procure a post long enough to stand 7½ or 8 feet out of the ground; if turned, with an ornament at the top, it will look all the better. Eighteen inches above the ground, set in six arms to support a rim four inches deep, and ten feet in circumference; halve the ends of the arms on to the under side of the rim, and fasten with nails or screws. Three feet above this rim, put another just like it; put in some

eyes made of wire, at the top of the post, say twelve or fourteen. Divide the rims into as many spaces as you have put in eyes, and stretch some No. 16 galvanized wire from the eye round a nail in the edge of the top one, and fasten it securely at the bottom. Give it one or two coats of paint, and it is ready for the vine.

Plant two vines under or near the trellis and grow them with double arms. Train the arms, one pair around the upper and one around the lower rim. Allow two fruit canes, after the first year of fruiting, to each wire; keep them tied to the wires, and by midsummer the trellis will be covered and will look very pretty, especially if you get near enough to see the rich clusters of fruit. This gives the same amount of vine as on a straight trellis ten feet long and two tiers high, and it can be used in many places where other kinds can not. (1869)

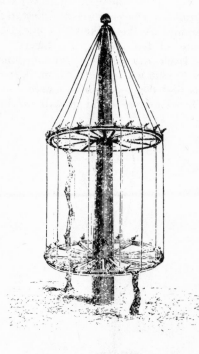

Wine-Making Suggestions

In making wine there are so many little details to be observed and so many things requisite to the best success, that it is not possible to give more than the most general directions. Wine-making is a trade which has to be learned either by one's own experience or from that of others. The quality of wine is affected not only by the process of manufacture but by the variety of grapes; and again the same kind of grape will produce a different product on different soils. Upon large estates in Europe, celebrated for their wines, vines in different parts of the same vineyard produce wines of very different qualities.

Then there is a great difference of opinion as to what constitutes wine. Some apply the term to grape juice fermented with the addition of sugar and afterward fortified with a portion of brandy or other distilled spirits. We consider none of these compounds as wines. The only thing which should be called wine is produced by the fermentation of pure grape juice without any additives whatever. Many of our native grapes will not make a wine that will keep, yet these differ according to the locality. Thus: the Concord is a valuable wine grape in Missouri, while many at the East say that wine can not be made from it. Sugar is generally added to the juice of the Isabella, yet we recently tasted, at the house of a friend who would not deceive us, a very good, very light wine made from the pure juice of the Isabella.

Whatever the variety of the grape, it should always be left upon the vine until thoroughly ripe. A few light frosts will do no hurt, and unless the grapes commence to decay they are better left on until there is danger from frost. The fruit is to be picked carefully, all imperfect berries removed in picking from the stems, and bruised without crushing the seeds. The bruising may be done in a barrel with a pounder, or they may be run through a mill for the purpose.

After grapes have been crushed, the further treatment varies. The juice which runs from the bruised grapes may be taken for the best kind of wine, and what can be pressed from them for a second quality, or the whole may be mixed together. The grapes must be pressed; this is usually done with a screw press, the bruised fruit being put in a coarse bag. If a light colored wine is desired, the grapes are pressed soon after they are bruised, but for a dark wine, the bruised grapes are put into a covered tub in a cool cellar and allowed to ferment. When the mass of pulp and skins rises to the top, and this crust begins to crack from the escape of bubbles of gas, then the pressing takes place. The time allowed for this fermentation on the skins will determine in a great measure the quality of the wine. The longer it is allowed to continue, the higher colored and the rougher, or more astringent the product will be.

In whichever of the above methods the juice, or *must*, is obtained, it has to be fermented. For this purpose it is put into a perfectly clean cask. A bung is then fitted to the cask which has a bent tube inserted in it. This tube is bent like an inverted letter U, one leg of which is inserted into the bung and the other dips into water placed in a cup or other vessel. By means of this arrangement, all the gas liberated during the fermentation passes out through the water, while the air is prevented from coming in contact with the liquid in the cask. The fermentation commences in a day or two and continues for several weeks. The lower the temperature the slower it will go on, and the better the quality of the wine will be. When bubbles cease to pass through the water in which the tube is immersed, remove the bung containing the tube, fill up the cask with juice which has been reserved for the purpose, and place a sound bung in lightly. A month later the bung may be driven tight.

Some time during the winter the wine is carefully drawn off from the lees into another perfectly sweet cask. In the spring, about the time of the blossoming of the grape, another fermentation takes place, at which time the bungs should be loosened. After this is over, the wine will usually become clear without any aid, and in a few months may be bottled, though the operation is usually deferred until winter. This is a mere outline of the process, which is variously modified according to the kind of wine desired and the peculiar views of the maker. It is essential to use the ripest grapes, observe the greatest cleanliness in all the vessels used, and to keep the casks full in order that the air shall come in contact with the new wine as little as possible. All the wines made from our native grapes, without addition of sugar or alcohol, are very light and will not bear exposure to the air. (1864)

Sundry Notes on the Cultivation of Hops

The staminate or barren flowers, and the pistillate or fertile flowers of the hop are borne upon distinct plants, and cultivators distinguish them by the terms male and female. The two plants are only to be recognized with certainty when in flower. The male flowers are in loose cluster springing from the axils of the leaves, as in fig. 2, which represents the upper portion of a flowering branch. A separate flower, consisting of a calyx of 5 sepals and as many erect stamens, is also shown. The pistillate flowers, which are borne in little cone-like clusters, are very simple in structure, and consist of a scale-like calyx, with the pistil at its base. After fertilization, the clusters of the pistillate flowers increase very much in size and become hops, fig. 3, which consist mainly of the enlarged scales of the calyx, each with a little nut at the base. Near the lower part of the scales are numerous oblong, resinous grains, called lupulin. One of these, very much magnified, is shown at the lower

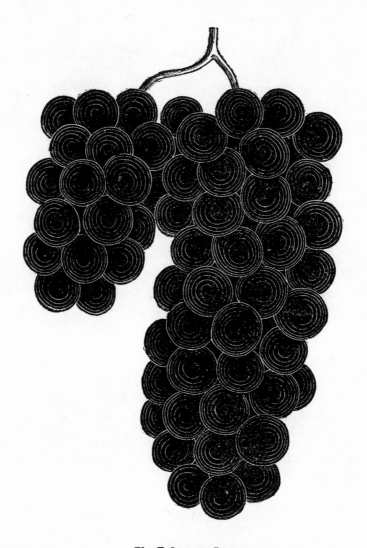

The Delaware Grape.

months of the year. It is already known to many of our readers that this city, New York, is greatly indebted for this luxury to several families by the name of Bergen, who annually cultivate some hundred acres, near Gowanus, Long Island, and at Shrewsbury, New Jersey. Although not a sure crop, we have been informed that an acre of their land, well tilled, will yield from $100 to $400 worth of melons in a season.

right corner of fig. 3. It is upon the lupulin that the valuable properties of hops mainly depend; hence in the collection and management of them, care should be taken to lose as little as possible of this. Lupulin is often incorrectly spoken of as the pollen of the hops. This is a great mistake, as it appears only in the pistillate or female flowers, and long after they have been fertilized by the pollen of the staminate or male plants.

stakes are 9 feet long, about as large as bean poles, and are set 1 foot in the ground. The horizontal hop yard has received the attention of many of the best hop growers of Central New York, and from what we can learn it possesses very important advantages: economy, early ripening, greater yield, less labor, less peril from wind, less shade, and avoidance of bleeding. The last specification is particularly noteworthy for great damage

BAY VIEW MELON.

The soil best suited for the melon, in open culture, is a light, sandy loam, similar to that of the southerly end of Long Island and the adjacent shores of New Jersey. The ground should be plowed or spaded, from 12 to 18 inches deep, and well pulverized with a harrow or rake. The proper season for sowing is at the time the peach tree is in bloom; for, if planted earlier, there would be fear of their being cut off by frosts. The seeds may be sown in broad hills, 18 inches in diameter, and 5 feet apart from center to center, each supplied with a shovelful of well-rotted stable or barn yard manure. In order to guard against accidents, at least 20 seeds should be scattered in a hill, which should be covered with finely-pulverized earth at about the same depth as in planting Indian corn.

Soon after the plants are up, and begin to show their second leaves, they may be weeded with a hoe, and a portion of them thinned out, still leaving enough to guard against accidents or the depredation of worms. In the course of the summer, before the vines begin to spread, two furrows should

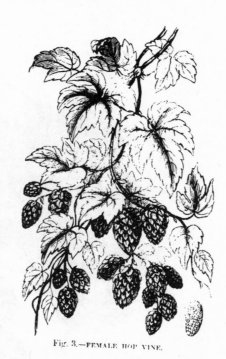

Fig. 2.—MALE HOP-VINE.

Fig. 3.—FEMALE HOP VINE.

Hops, like corn, grow on all varieties of soil, from swamp muck to gravel knolls, and any good corn land is suitable for them. In preparing the ground, put it in good condition and plant corn, potatoes, beans or some similar hoed crop. Leave places for the hop hills 8 feet apart each way in the rows.

A plan has been invented and patented by Mr. F. W. Collins, of Morris, Otsego Co., N.Y., which obviates many difficulties heretofore encountered, and saves 75 per cent of the expense of poles, etc. This is shown in the engraving, fig. 1. The poles or

is annually done to the roots by cutting off the vines near the ground at harvest, as is uniformly done in order to raise and remove the poles with their burden of fruit. (1864)

Cultivation of Melons

There are many varieties of the melon (Cucumis melo), and of these, several abound in our markets for at least three

Phinney's Early Water Melon.

146

81-POUND CUBAN QUEEN WATER-MELON.

be run between the rows, with a cultivator or plow, turning the earth directly from the plants, which should again be freed of weeds, and reduced in number to five or six in each hill. A few weeks later, a second plowing should take place, turning the earth towards the vines, when a broad, flat hill should be formed, slightly hollowing in the middle, so as to receive and retain the water supplied by irrigation or from the fall of rains. After this, no further attention will be required, except in keeping down the weeds, and in guarding against worms. (1848)

Green Citron Melon.

Protecting Melon & Other Vines

Every one at all conversant with raising melons, cucumbers, or squashes, knows well the propensity of the bugs to eat both leaves and stalks. Some years they are worse than in others, and it is necessary to be always prepared for them, and this is best done by preventing their reaching the growing vine with some sort of screen or covering.

Our usual practice, as soon as the vines appear, is to cut old newspapers into pieces about 20 inches square, and place one over each hill, covering the border with earth. This effectually keeps off the bugs. The plants raise the paper as they grow. But during a rain or heavy dew the paper breaks and the plants are at the mercy of the bugs. This plan does not answer well for the larger squash or pumpkin plants.

Fig. 1.—BOX PROTECTOR.

A good protector is shown in fig. 1, consisting of a box 12 inches square, and 5 or 6 inches high, with a piece of mosquito netting nailed acorss the top. Wire netting is more durable, and if painted such "boxes" will last a lifetime. Make the boxes bevelling, that is, wider at the bottom, for close packing in nests when not in use. An effective and cheap plan is to soak mosquito netting in oil, and when dry place it upon a frame over the vines. One method of doing this is shown in

Figs. 2 and 3.—WIRE SUPPORTERS.

fig. 2. The ends of two bent wires are placed in the earth to support the netting. A modification of this plan is shown in fig. 3. Still another plan is given in fig. 4, in which the supports are of wood. The netting should be cut about 24 inches square, and when placed over the frame, cover the edges with earth. Vines thus protected are safe, and will attain quite a growth before the covering need be removed. (1882)

Fig. 4.—WOOD SUPPORTER.

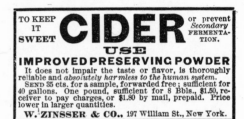

XI Wood & Trees

Clearing Forest Lands

As we resided upward of seven years in the western country, part of the time hutting it in a log cabin, and nearly the whole of it actively engaged in clearing forest lands, and bringing them into cultivation, we profess to know something practically of the subject upon which we are about to write.

There are several methods of clearing land, dependant entirely upon the price of wood and timber in the neighborhood of their location. Where these are valueless, except for the purpose of fencing the farm and making its buildings, the slashing system of clearing is usually resorted to. This consists in taking a very large tree for the center, and cutting it off as nearly as is possible to do so without endangering its falling. Then begin and cut all the trees in a circle of a hundred feet or so from this center, the same as the first, and in such a manner as to ensure their falling toward the center one if possible. When the circle is thus cut, four men with their axes take opposite sides on the outside of the circle at the largest trees, and commence cutting on them till they fall, taking care that the four trees shall come to the ground nearly as possible at the same time. These, in falling, generally carry all the other trees with them that they touch, and they again others; so that when the circle so cut has finished falling, nearly all the trees in this diameter of 200 feet will lie with their heads pointing to the center, their butts out, and lapping each other like reversed shingles.

Others cut the trees so as to fall in a line head to head, making them appear, when they get through chopping a swath, as if they had been raked into winrows. To lay the trees in circles or rows, requires both judgment and address; and none but experienced hands should undertake it, otherwise they will

fail entirely in accomplishing their task. Slashing is also performed by cutting the trees and allowing them to fall as they please; but this is considered very unworkmanlike, and unless the wood and ground are very dry at the time of firing, it leaves many more logs to roll up into heaps for burning than either of the plans first mentioned.

Heavy log-rolling is not only an expensive, but an excessively laborious business, as we well know to our cost. We reckon it equivalent to digging rocks and laying stone wall; a business also in which we profess a trifle of experience. The extra number of logs usually left to be rolled after firing, let the trees be cut with all the address possible, is the greatest objection we have heard to the slashing system of clearing land, and although it saves labor in the first operation, unless the soil be a dry one, we would not recommend its being adopted. Where the soil abounds with clay, the land is tenacious in holding water, and if the forests growing upon it are slashed, the trunks of the fallen trees get so saturated with moisture that scarce one summer in four will prove sufficiently dry to burn them. It will be necessary to let the trees slashed lie till the second summer after being cut, before they will effectually burn.

The second method of clearing land is, as fast as the trees are fallen, to cut them up into logs of convenient length, say 15 to 30 feet long, and roll them together, placing the largest at the bottom, and then pile the tops of the trees upon them, and burn when the wood becomes thoroughly dry.

The third method is to clear woodland for pastures. This consists simply in underbrushing the forest, and cutting out all the small growth, and such other trees as are likely to be prostrated in a high wind, and piling them in heaps and burning when dry and then sowing grass seed and harrowing it in. Woodland pastures answer very well in the latitude of 40 degrees, and south, north of 40 degrees, the summers are usually too short for reserving woodland pastures to any great advantage, except the soil be calcareous, and blue grass comes in naturally.

Where wood and timber are valuable, the method we adopted in clearing the land, was first to complete underbrush and pile it; cut out the wood, split and pile that; and lastly, the timber, cutting up the tops of the timber trees if not suitable for wood, and piling them. After removing the wood and timber, and the top heaps and brush became dry, we burnt, thus leaving the land clean and fit for a crop.

We have often wondered at the almost total absence of taste in clearing lands displayed by our countrymen. They usually commence on the line of the road and take a clean sweep through their farms to the back of them, where they make a small reservation of wood and timber for use. This leaves the land completely unprotected from the fervid sun in summer, and the cold searching wind in winter; and then the horrid stumps stand out in bold relief, staring one in the face with their black charred, or rotten punky sides, for half the age of man; making one of the most desolate and dreary sights we ever looked at. Nearly all this may be obviated, and even a new country recently cleared be made to assume a handsome, cheerful appearance, by proper reservations of the original forest. Suppose the farm 100 acres, we would divide it into five 20-acre fields; or if the size of 200 acres, into 40-acre lots, and in clearing, reserve a belt of trees around them from three to six rods wide.

The advantage of these would be threefold: 1. They would afford all the wood and timber necessary for the future use of the farm. 2. Shelter it from the rude blasts in winter, and give it shade for the stock in summer. 3. Hide the stumps and give a handsome, picturesque appearance to the country.

An occasional group of trees left near the center of the field, especially where the land was highest, would also add to the beauty and variety of the farm. The only objection we have ever heard to such reservations were, that in narrow belts and small groups, the trees were apt to be blown down. This danger may be obviated by cutting out all the tall trees, leaving the shorter and younger ones only. Left in this manner, they shade and protect each other, spread out their branches and roots as they grow, and soon strike the latter so deep into the ground, as to enable them to stand against the strongest winds. Some of the handsomest farms in Europe are thus laid out into square fields, with belts of trees around them for shade and shelter, and the profit of the wood and timber. In order to obtain these the owners have been at great expense in planting them. We have only to leave what nature has already prepared to our hand, to equal these in picturesque beauty. We wish that the owners of wild lands would consider this subject, the sight of one such farm could not but convince them of the real utility, if no other motive were wanting of adopting this course of reserving belts of trees around the fields in clearing.

May, June, and July are the best months for cutting forests; but this is a time that the farmer can ill spare for such work; necessity, therefore, compels him to do his chopping in the winter when he has little other employment. Firing the log-heaps should take place in dry weather, and when a gentle breeze prevails. (1843)

When to Cut Timber

Conversing with an intelligent farmer of large experience, upon this subject, we found he fully sustained the view heretofore expressed: that the best season for cutting timber is about mid-summer. His explanation was, that during the latter part of June and early in July, when the foliage is in its fullest vigor, the upward draft upon the sap is so great that very little moisture is left in the tree, consequently the timber seasons hard and sound; but that during March and April there is so much water in the wood, that insects bore into it readily, thus producing "powder post" through all the sap portion, and even into the heart wood. He mentioned the instance of a neighbor who cut his timber for a house in June, but when he came to work it out in the winter, he lacked some ribs or slats upon which to nail the long roof shingles. He cut enough to supply the deficiency during the latter part of the winter, and completed the house. After the lapse of a few years, he examined the roof, and found the slats which were cut in summer, perfectly sound, while those cut in winter, were badly affected by dry rot and "powder post." Our informant, it is sad to report, had also proved the same thing himself. He also remarked that when the object is to induce a free growth of new shoots for a future wood or forest, he preferred to cut in March, as the stumps sucker much more freely then, than when cut away in summer. The latter, however, is the best season to clear off a growth of wood; the old stumps decay soonest. (1860)

The Wood-Lot in Winter

A few acres in trees is one of the most valuable of a farmer's possessions; yet no part of the farm is so mistreated, if not utterly neglected. Aside from the fuel the wood-lot affords, it is both a great saving and a great convenience to have a stick of ash, oak, or hickory on hand, to repair a breakdown, or to build some kind of rack or other appliance. As a general thing, such timber as one needs is cut off, without any reference to what is left. By a proper selection in cutting, and the encouragement of the young growth, the wood-lot will not only continue to give a supply indefinitely, but even increase in value.

A beginning, and often the whole, of the improvement of the wood-lot, is usually to send a man or two to "brush it," or clean away the undergrowth. This is a great mistake. The average laborer will cut down everything; fine young trees, five or six years old, go into the heap with young poplars and the soft underbrush. The first point in the management of the wood-lot is to provide for its continuance, and generally there are young trees in abundance, ready to grow on as soon as given a chance. In the bracing winter mornings one can find no more genial and profitable exercise than in the wood-lot. Hardwooded and useful young trees should not have to struggle with a mass of useless brush, and judicious clearing up may well be the first step.

In timber, we need a clean, straight, gradually tapering and thoroughly sound trunk. In the dense forest nature provides this. The trees are so crowded that they grow only at the upper branches. The lower branches, while young, are starved out and soon perish, the wounds soon healing over are out of sight. In our open wood-lots, the trees have often large heads and the growth that should be forming the trunk is scattered over a great number of useless branches. Only general rules can be given in pruning neglected timber-trees; the naked trunk, according to age, should be from one-third to one-half the whole height of the tree; hence some of the lower branches may need to be cut away. All the branches are to be so shortened in or cut back as to give the head an oval or egg-shaped outline. This may sometimes remove half of the head, but its good effects will be seen in a few years. In removing branches, leave no projecting stub on the timber, and cover all large wounds with coal tar. Whosoever works in this manner thoughtfully can not go far astray. (1883)

CEDAR OF LEBANON, AT WOODLAWN, NEAR PRINCETON, N. J.—Hight 36 feet.

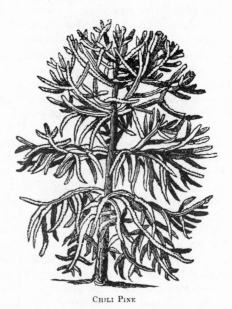

CHILI PINE

The Chili Pine

The Chili pine, or *Araucaria imbricata*, is a tender tree, and needs not only the protection of shade, but of the dry sub-soil. The accompanying figure is a portrait of a young tree, 12 feet high. On its native mountains (the Cordilleras,) it attains a height of 150 feet. (1873)

Cedar of Lebanon

This magnificent tree, to which the Scriptures make very frequent allusions, derives its name from Mount Lebanon, in the vicinity of which it most largely abounded formerly. It is often spoken of in connection with buildings, and particular reference is made to Solomon's Temple, and the fourscore thousand (80,000) hewers in the mountains preparing the timber. Some writers think that the inroads made upon these forests at that time so thinned them out that they have never fully recovered. Certain it is, that late travelers do not find them abundant in those regions, although a few very large specimens are left, whose ages must date far back into the past, as many of them now measure over 30 feet in circumference.

The trees were introduced into England many years ago, and succeed well in that moist atmosphere. One is described by Loudon, 72 feet in height and 24 feet in circumference. Another is spoken of which was blown down in 1779, and measured 70 feet in height. It is a rapid growing tree in that country, after the first few years. They are not perfectly hardy in this country north of 40 degrees, but in most localities south of that parallel succeed well. We have seen several beautiful specimens at Flushing, Long Island, of some 40 or 50 feet in height, with their broad, depending branches sweeping the ground in a circle of about 45 feet in diameter. They show best when grown as single trees; the lower branches die out when they are crowded together in masses.

They are produced, with some difficulty, from cuttings, in propagating frames or houses, but are more generally raised from seed, sown in Spring upon a rather light sandy loam, covering only one-half inch. They will require a slight protection in this latitude, for the first few winters, after which, plant in a deep soil, somewhat moist. When grown in perfection, they form a splendid tree, as shown in the engraving. (1859)

Fig. 3—FOUNTAIN PINE—OR WIDE-SPREADING MEXICAN PINE (*Pinus patula*)—Hight about 5 feet.

The Beech Tree

A Western artist was commissioned to prepare some original sketches of native forest trees, and among others we received the accompanying group of beeches *(Fagus)*. Though beautiful as a picture, the sketch does not give a full representation of the peculiar characteristics of the beech tree as it grows wild, thickly studding many extensive forests that abound on the fertile soils of the Northern and Western states. As we have usually seen it, it has a straight, tall trunk, the bark smooth, with scattering small branches shooting out in every direction, beginning at a height of from 5 to 30 feet from the ground. In the forest the main branches are usually from 15 to 20 and sometimes 40 feet from the ground. Wherever this tree abounds, it is a favorite one with the new settler for the construction of log houses.

It bears an abundance of fruit, beechnuts, of triangular shape, or three-sided, and not unlike buckwheat kernels in form and even in color, though the beechnut is of course much larger. (The name *buckwheat* is derived from beechnut, or beech-wheat, from its resemblance to the nut.) Beechnuts, called *mast*, or *beech-mast*, are very nutritious and serve to fatten a vast number of hogs in the newer countries. Hogs and other animals thrive well upon them, but two or three weeks of final feeding upon corn are required to give solidity to the pork, though we have assisted in slaughtering many hogs immediately after driving them in from the forest, when further feeding was shut out by deep snows, and the pork was found to be delicate and good for home use, though not well adapted for strong salting for distant markets.

The nuts are pleasant eating, and in our boyhood days a bushel or two of beechnuts were considered no mean addition to the garret stores, especially when the hickory nuts and black walnuts chanced to yield poorly. The nut grows in a bur which opens and drops out the kernels after a severe frost, while the burs still clinging to the trees present a pretty appearance in winter. The wood of the beech ranks next to the oak and maple for fuel. It is of close texture but unfitted for timber in exposed situations, owing to its liability to decay. It is admirably fitted for many mechanical purposes, and is much used for making planes, shoe-lasts, saw and other tool handles, wooden screws, rolling pins, butter stamps, etc. The beech has a dense foliage, and makes a pretty ornamental shade tree for standing singly upon the lawn, or in groups. The weeping beech *(Fagus pendula)*, with its long pendant branches hanging down nearly to the ground, is scarcely excelled in beauty by any other tree. There are several varieties of this tree—the red beech *(Fagus ferruginea)*, and the white beech *(Fagus sylvatica)* being the most common in the northern portions of our country. The former is thus described by Gray: Leaves oblong-ovate, taper-pointed, distinctly and often coarsely toothed: petioles and midrib soon nearly naked; prickles of the fruit recurved or spreading; common, especially northward, and along the Alleghanies southward. (1859)

Cultivating the Black Walnut

The demand for black walnut timber has so increased of late years, not only for home use, but for export, that the trees are rapidly disappearing. The high price of the wood, which will increase, rather than diminish, now allows it to be brought from great distances, and the natural growth is disappearing wherever there are means of transportation. It is not surprising that the attention of many is turned to the cultivation of black walnut, and we receive letters from widely separated localities, inquiring as to the probable profits of such plantations, and asking for directions for establishing them.

As to the first point: there can be no doubt that, by the time plantings made now, afford timber of proper size, the price will have reached a point which will afford handsome returns. Even at the present price, landowners in Great Britain are contemplating making plantations of the trees. Several inquire if the tree will succeed in their locality, and these inquiries come from Mississippi, New York, and other states. The black walnut is remarkable for its wide range, it being found in Florida, Louisiana, and Texas, at the South, all the way to the New England and other Northern states, attaining its greatest development in the fertile soils of the Mississippi valley. The experience of the few who have made plantations of the tree, shows that in its early growth it needs the protection of other trees or "nurse trees," and that it must have cultivation to prevent the encroachment of grass. The tree does not transplant readily, and like others with a strong tap-root, succeeds better if the seed is planted where the tree is to remain.

As with other timber trees, the value of which consists in having a tall clear trunk, this must be planted thickly, to prevent the growth of heavy side branches, and thinned out as the trees become crowded. The land is marked out in rows four feet apart, and the nuts planted eight feet apart in every other row, the intermediate rows being devoted to potatoes or corn. This cultivation in the alternate rows, is continued for two years, when the spaces between the rows of walnuts are planted with nurse trees. The silver maple, the European larch, and even the white willow have been used. These are to be thinned out as they are found to encroach upon the walnuts, and are useful for poles, or may be employed as fuel. If any of the walnut trees are disposed to branch low, they should be pruned while small, to secure a tall, clear trunk. When the trees crowd one another, each alternate one must be cut out; though not large enough for lumber, they will serve for various home uses. The seed is planted in spring. It is best preserved during the winter by placing it in heaps of a few bushels, upon a dry spot, and covering with sods, upon which should be placed several inches of earth, as in covering pits of roots. (1883)

Evergreen Hedges

Visiting the estate of Mr. J. P. Cushing, at Watertown, near Boston, a few weeks ago, I was struck with the beauty and economy of some arbor vitae hedges enclosing his grounds. They have acquired a sufficient growth to prevent the passage of all ordinary animals, and have the appearance of enduring many years. It occurred to me that other evergreens, as the spruce and fir tribes, could be applied in the same manner, and serve for the same purposes. Although the arbor vitae possesses more beauty than the spruce, and fir, yet we conceive that this is counterbalanced by the advantages attending their cultivation. The latter are cheap, abundant, and will grow in almost every soil and climate.

The following diagram will show the mode of setting out these hedges.

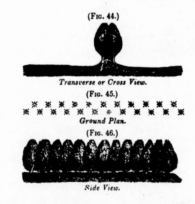

(Fig. 44.)

Transverse or Cross View.
(Fig. 45.)

Ground Plan.
(Fig. 46.)

Side View.

The young shrubs are planted alternately in two rows about nine or ten inches apart, as denoted in fig. 44, on a ridge of earth slightly elevated above the common level of the ground. The branches of one row entwine themselves among those of the other, and among themselves, forming a thick, tufted mass as indicated by fig. 46, which, in ornamental plantations, should constantly be kept clipped, in order to preserve uniformity.

When first set out, the young trees appear to be too far apart, leaving spaces wide enough to admit small animals; but, in the course of a few years, their trunks will increase in thickness, and form as complete a barrier as the thorn or holly. (1843)

The Maple Tree & Its Sugar

Maple sugar is mainly obtained from the sugar maple (*Acersaccharinum*), also called the Rock Maple, which grows chiefly in the Northern and Middle States east of the Mississippi River. The sugar maple is a most valuable tree, not only for its sugar product, but as fuel it approaches hickory, and is the best of all woods for charcoal. Its hardness and frequently curled grain admirably adapt it to cabinet work. It is also a beautiful shade tree for the street border, and for some fields and lawns. A sugar maple grove adds greatly to the value of any farm, and multitudes of farmers should plant one for their own enjoyment in later years, and for their children.

Saccharine matter is stored in the tree in the form of insoluble starch, which is changed to soluble sugar when the sap flows after the winter's frost, and is by it distributed to the twigs, and to the leaves when they develop, and to all parts of the tree where growth is to be made. The best flow of sap is on a warm day following a freezing night. Inserting a tube near the base arrests and draws off a portion of the circulating sap. This, caught in rude wooden troughs, or in pails or buckets, is boiled down sufficiently to drive off a large portion of the water. The syrup is left to cool, and the sugar crystallizes, differing from the Southern cane sugar only in its peculiarly pleasant flavor, which is almost universally liked, and gives it a ready sale.

Any boiling vessel will answer, from an iron pot or kettle on the stove, for a small quantity, up to the immense caldron, set on stone, or brick masonry, or hung on a pole, supported by stakes, with a fire built around it on the bare ground. The improved boilers are shallow pans having a large evaporating surface. Successive portions of fresh sap are added until there is sufficient concentrated to "sugar off." The fire is then slackened, and the syrup constantly stirred to prevent its burning. When so thick that a little of it, cooled on a spoon, or the end of a stick, takes a hard waxy form, a little brittle, the fire is removed, and the crystallizing takes place on cooling. If the sap has not been kept perfectly clean, the boiled syrup is trained through a thick linen cloth before the final concentration.

As ordinarily made, maple sugar is quite brown. If the flowing sap is collected in clean covered vessels, and no leaves, or dust, or other foreign substances are allowed to get into it through the entire process, the sugar will be almost as white as the common refined cane product. Half-inch augur holes in the trees, with galvanized-iron spouts to fit—one for small trees, and two or three for large trees—are much preferable to wooden spouts, as they injure the tree far less. (1883)

Maple Sugar

Immediate preparation should be made for the work of sugaring, particularly where there is a large sugar-grove. Provide a good supply of dry fuel convenient to the boiling place of the ssp. A shed to enclose the arch and kettle will add to comfort and cleanliness. Buckets of tin are lightest to carry, but are apt to be rusted during the year, which would impart a dark color to the syrup and sugar. Unpainted pails of cedar, pine or whitewood are generally preferred. These, with the evaporator, the spouts and everything connected with the manufacture, should be kept scrupulously clean. The need of clarifying syrup, arises mainly from neglect in this respect; the best specimens we have seen were made without the use of any substance to remove impurities—none were suffered to be mixed with the sap.

Spouts are readily made by removing the pith from pieces of elder, or from foot lengths of inch square pine. For the latter, remove the upper half to within two inches of one end, bore or burn a one-quarter inch hole through the thick part left, and cut a groove from the hole to the other end, as in fig. 1. If

Fig. 1.

elder stalks are accessible, good spouts are made as illustrated in fig. 2, by sawing half through at *a* and *b*, and splitting between the cuts. Each piece then makes two spouts.

Fig. 2.

A three-quarter inch auger bit is best for tapping, which should be done on the south side of the tree, boring the hold about one inch deep.

It saves time after the boiling is commenced, to have a constant stream of sap trickling into the evaporator, and the thickened syrup discharging into a second vessel for "sugaring off." Otherwise, the sap collected must wait until the first lot is finished. In the latter part of the season it readily sours, and may soon spoil. It is well to add a little lime to the sap during the last running, to neutralize any existing acid. (1863)

Maple Sugar Making

Wooden sap troughs and potash kettles are still in use in some parts of the country, but enterprising sugar makers use wooden buckets which are preferable to tin, and flat evaporating pans, and the sugar is much improved. The sap is sometimes conducted to the sugar house in "leaders" or small wooden troughs, which would be improved by scalding them out once a day to prevent souring. In like manner the buckets ought to be scalded occasionally.

The trees are tapped with half-inch augurs, and the hole enlarged with a sixteenth of an inch larger bit, before the close of the flowing season. The sap spouts are 6 or 8 inches in length, 1-inch square, or turned round having a 1-inch hole bored through them. The ends are tapered off, and they are driven into the holes of the trees so as to barely hold. If tubs are used to collect the sap, there should be holes of about 10 inches square cut to pour in the sap, and over them linen towels should be laid, to strain out sticks, leaves, etc., if the arrangements of the buckets, etc.,

are not so perfect as to exclude all filth, as is desirable.

After this, the sap must be kept covered. The storing tubs should stand on higher ground than the boiling pan, so that the sap will flow from one to the other. During the boiling, skim as often as scum arises. It is seldom that much skimming is necessary. When the cooled syrup is nearly as thick as good molasses, draw it off into a tub to settle, straining through a flannel strainer. Here any sediment will be deposited. After the syrup has settled clear, draw it off, and boil it down again until it is thick enough to sugar off. When the sugar is to be "caked" or "stirred," it must be boiled until a spoonful of it put upon snow will be perfectly brittle when cold. The liquid sugar is taken from the fire and when granulation has commenced, and the mass is thickened considerably, fill the moulds rapidly. If it is to be stirred, at the same time commence stirring, the kettle being held firmly, and stir the mass till it has the appearance of dry brown sugar of the shops. When the sugar is to be drained it is usually taken from the fire before it granulates quite thoroughly, it is ladled out into tubs with false bottoms, some 5 inches above the true, 3 or 4 holes being in the false bottom, and covered by saucers or plugged by round smooth sticks. The sugar is ladled into the tubs, and when settled the plugs are loosened and partly withdrawn, so that the molasses will run through. This may be drawn off from the bottom of the tubs. (1864)

Maple Sugar Making

The use of metallic spouts for tapping sugar maples has become nearly universal, and with good reasons. They are now obtainable at a very low price. Adhering firmly to the bark, they make a secure support for the bucket, without necessarily being driven so hard as to shut off the pores of the sap wood, and they allow the use of a small bit for tapping. They are more easily kept clean and sweet than wooden spouts, and being shaped by machinery are of uniform size and taper.

I bore with a three-eighth inch bit, and use a plain tin spout. Being free from obstruction inside, they allow the use of a small swab in cleaning, which should be attended to as scrupulously as the cleaning of the buckets. In thrifty trees bark will grow over the wound in one summer, greatly diminishing the chances of decay. By using a hard wood mallet for driving, and by exercising reasonable care, they will last many years.

I open first with a three-eighth bit, boring about three-quarters of an inch deep, inserting the tin spout to support the bucket. The spout is not afterwards removed during the season. At the end of the first heavy run I insert a wooden spout four to six inches long, above and to one side of the first boring.

These spouts I make of second growth **white ash** sprouts cut between joints, and the pitt burned out with red hot No. 10 or 12 wire. I use a foot lathe to taper the ends. They do not need hard driving, as they have nothing to support.

We are sometimes advised to tap on the south and east side of the tree to secure an early and abundant flow. While the reasons given sound plausible, I think they possess no practical value. When the weather is too cold to start sap on all sides of the tree, the flow will not be sufficient to be profitable. When tapping, I look only for a good place without regard to location, except to get some distance from last season's scars, usually as low down as is convenient in gathering sap.

I use wooden buckets exclusively. My method in recent years at the last gathering is to have the buckets brought near the roadway and placed in bunches right side up. As soon as this is done we fill gathering tanks with water, go back over the camp and fill the buckets. They are then allowed to stand and soak a day or two. We then go around and scrub them thoroughly, and turn them over to drain. When dry they are packed away bottom up, and they are ready for use again. Bucket covers of wood or tin are rapidly growing in favor. They keep out flying leaves, moss, and dead bark, insects to a considerable extent, and, what is more important, rain and snow. (1888)

Holder for Splitting Wood

Mr. A. R. Dixon, Lake Co., Ohio, has a simple and effective holder for use in splitting wood, which is shown in the accompanying drawing. A crotch is cut from a tree a foot in diameter, and a piece of heavy plank is nailed upon the two ends by large spikes, thus making an enclosed space in which the wood is placed while being split. Mr. D. writes that when a lad he split a great toe with an axe, and hopes this device, which has been of so much good service to him, will prevent others from suffering a like injury. (1882)

A WOOD HOLDER.

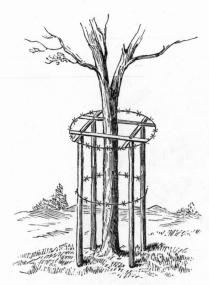

A TREE PROTECTOR.

Tree Protector

A good fence, and gates that horses can not open, make the best protection for trees. Our correspondent, "G.E.B.," Litchfield County, Conn., while aware of this, finds it necessary to sometimes plant trees where animals must run for awhile. In such cases he protects each tree with a barrier of its own. He drives down four stout stakes, equally distant from the tree. Upon the tops of the stakes are nailed four strips, the ends of which project several inches beyond them. A piece of barbed fence wire is then nailed to the ends of the projecting strips (probably best put on with staples) in the form of a circle, as shown in the engraving made from the sketch sent us by Mr. B. Another piece of the barbed wire may be put around the stakes, lower down, to prevent the animals from reaching below the first circle and gnawing the trees. Our correspondent says the animals soon learn to respect this barrier and leave the trees alone. (1882)

Wood Rack & Wood Apron

The engraving shows two convenient methods of carrying fire wood. A wood rack for the shoulder is made of a piece of round hardwood, with four long pins set in the upper side. These pins are placed in V-shaped pairs, between which the fire wood is piled. A handle, 3 feet long, is set in a hole bored in the center of the underside of the body of the rack. This device, when complete, resembles a "skeleton" hod, and is carried in the same manner as a hod for brick or mortar. A second method of carrying wood consists of a stout canvas "apron," in the lower part of which the fuel is placed, as shown in the engraving. A boy, or other person, with much

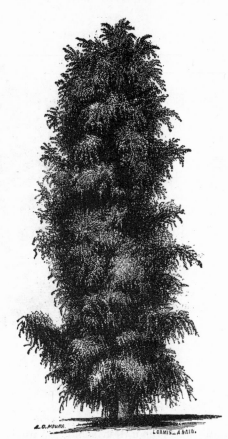

Fig. 5—WEEPING JUNIPER, (*Juniperus pendula*,) Hight 6 feet.

wood to move short distances, will find either of these methods labor-saving, and they are quickly made. (1883)

WOOD RACK AND APRON.

156

A.Q.MOORE.DEL.   J.TRENT.SC

Fig. 1—EUROPEAN LINDEN, (*Tilia Europæa*)—Hight 78 feet.

XII   Wind & Water

Water Heater / Japan Bath

A correspondent in writing from near Lucknow, in the East Indies, sends a sketch of what is in common use there as a Japan Bath, though it is employed to heat water for many purposes besides bathing. He thinks that such an affair would be much used in this country if it were generally known how quickly and cheaply water can be heated in it:

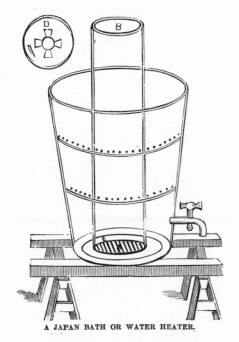

A JAPAN BATH OR WATER HEATER.

The reservoir used is a half-barrel, large firkin, or some such affair. In the center is a cylinder of sheet iron, B, which fits over a hole in the bottom of the vessel; it has a flange, E, at the bottom, by means of which it is fastened to the bottom of the barrel, by rivets, with red-lead between the two to make a close joint; there is a grate, A, at the bottom of the cylinder, to hold the charcoal. At the top is a cover, D, with openings that may be closed more or less, and serve as a damper. A faucet near the bottom completes the heater. It should be set upon two horses or other supports, placed far enough apart to clear the grate and allow the coals that may drop from it to fall to the ground. Our correspondent states that, properly managed, two lbs. of charcoal is enough fuel for a day. (1882)

Keeping Ice

There are few farms so small that it will not pay to provide for a supply of ice; and it is quite practicable for several small families to combine, and store a lot to be drawn upon by each. A fresh water lake, pond, or stream is not indispensable for obtaining ice.

In the absence of natural resources, an artificial bed can be made by excavating the soil a foot or less, or by enclosing an area with a small bank of earth. A thin grouting of hydraulic lime, spread thinly over the bottom and against the sides will hold the water until it freezes an ice flooring. Small additions of water daily or oftener in cold weather will soon give a thick mass of ice. Straw may be thrown over during a hot mid-day, or a temporary thaw. This water may be drawn from cisterns, or wells. Every four small hogsheads of water (63 gallons each), will give over a ton of ice, or 2,105 lbs. Even on the prairies, there are generally frequent sloughs which can be utilized for an ice pond with little trouble and expense.

A CHEAP ICE-HOUSE ORNAMENTED.

An ice house that will keep a large supply of ice is a small matter. We recently described a very good one built by a farmer (C. H. Warner of Lima, N.Y.). A New Jersey subscriber sends us a sketch of one very similar, indeed patterned after Mr. Warner's, of which we present an engraving herewith. It stands on level ground, and he has planted shrubby trees around it, partly for shade and partly for ornament. Some of these are shown. This house is 12 feet square, with sills and plates 8x8 inches, of hewn logs, and 8x8 corner posts, 8 feet high. Studding is set in as needed. Rough (or planed) boards are nailed horizontally within and perpendicularly without, and the cracks battened with narrow strips. The wall space is filled with sawdust. The roof is of single boards with a

160

ventilating opening at the top. The doors are single, with short cross boards inside to hold the ice up. The ice is packed in solid, except a space of 6 or 8 inches all around filled with sawdust. When full, a foot or so of sawdust is put on top of the ice. The flooring is of inch boards laid on a bed of cobble stones. (1882)

Cisterns Cheaply Made

About the majority of country and village houses, one sees old barrels with boards or troughs, and occasionally eave-spouts of tin, arranged to conduct water into them from the roof. The supply thus obtained is often inadequate to the wants of the family, and soon becomes unfit for use. Dead leaves, worms, all kinds of flying refuse, collect in the barrels, and they afford breeding places for myriads of mosquitoes; besides the supply of soft water is in a condition to use only a short time after a shower. Every house should have a cistern. None but those who have been accustomed to the use of soft water in domestic operations, know how much superior it is to hard. It not only makes things cleaner, but the labor of cleaning is greatly lessened, while hard water ruins woolens.

The expense of building cisterns is not large. In localities where the soil is composed primarily of clay, I have seen them built by digging a hole in the ground, and plastering the exposed surface with water-lime. Generally a layer of stones is placed in the bottom, over which a thin mixture of lime and water is poured. This runs down among the stones, fills all the crevices, and settles into a smooth surface above them, and, when dry, forms a floor of suffucient thickness and hardness, to allow any one to tread upon it while cleaning out the cisterns, without danger of breaking the floor. Plaster the lime directly upon the slanting sides of earth, with a common trowel. Care should be taken to not mix up a great quantity at any one time, as the lime soon "sets," and it becomes impossible to smooth the walls properly, if too large a surface has been plastered over roughly, before beginning the smoothing process.

The top of the cistern should be well covered, and pains taken to see that the earth on which the wall is put, is below frost in winter, otherwise freezing is likely to crack the coat of lime, and cleave it off, or leave cracks, through which the water will soak out. If properly banked up in winter, such a cistern will last for years; it is quite as satisfactory, indeed, as one built up of brick or stone. An opening large enough to admit a person should be left in the covering. In cleaning it, the walls can be readily washed, and all slime completely removed by using a stiff scrubbing brush. (1882)

A Filter for Cistern

Fig. 1 shows the filter. It is a large barrel with one end knocked out. At the bottom is a layer of fine charcoal, d. Above this is a layer of fine gravel, c; over this is a layer of coarse gravel, b, on the top of the barrel is a thin strainer, a, held in place by a hoop which fits over the barrel. The cloth is depressed in the center as seen in the cut. This strainer catches all leaves and coarse dirt, and should be cleaned after every shower. Some use a wire

Fig. 1.—SECTION OF FILTER.

strainer of very fine mesh, but the cloth answers the purpose very well. The filtered water flows through a hole, e. Into this hole a metal tube a foot or more long, punched full of holes, and covered with wire netting, is inserted. Six inches below the top is another hole, s, which is fitted with a short pipe as seen in fig. 2. During a heavy shower the overflow runs out of this hole, and into a spout provided for it.

Fig 2.—ARRANGEMENT OF FILTER AND CISTERN.

Fig. 2 shows the general arrangement of the entire apparatus. The barrel has a small shed built over it, to protect it from the sun and weather. This shed should be open at one end, so the barrel can be taken out at any time. The top is movable to allow the strainer to be cleaned. The lower section of the waterspout should be loose, so that it may be moved up or down, and turned. In fig. 2 the elbow rests on a block, or bracket, and the water flows through a hole in the cover of the shed, into the barrel. When the cistern is full, the elbow is turned, and drops down to a block, and throws the water into the spout to be carried away, or into a "wash water" cistern near by. The above arrangement may be modified to suit different circumstances and places. When rain water is used exclusively for cooking and drinking, it is best

to have a cistern for it alone, and a separate one for wash water. At the beginning of a storm, it is well to let the rain wash the roof for an hour or two, before the stream is allowed to enter the cistern. This is especially necessary where pigeons and other birds collect upon the roofs, as well as to wash off accumulated dust. (1882)

Water-Gaps/Water-Gates

When fordable streams cross highways, or through fenced pasture grounds, any contrivance which will let the water pass in time of freshets, without washing away, and yet form a good fence when the stream is fordable, is called a "water-gap," or "water-gate." This may be arranged to float upon the rising tide, or being stationary, let the water through or over it. The floating gates must be so constructed as neither to be broken by ice nor to entangle brush or floating logs and trees; fixed ones can only be used where much ice and flood-wood do not occur.

The first form which we give, fig. 1, is very simple, but faulty inasmuch as ice and snags would be very likely to catch in it. Very similar to this, is one without these defects, a sketch and description of which were forwarded by Adam Haun, of Washington Co., Ill. It consists, fig. 2, of two uprights, crotched at

the top, very firmly set in the ground, and braced against the direction of the flow of the water. Between these, and lying in the crotch, is a pole, larger or smaller according to the width of the stream. Near each end a short section is worked down to a smaller diameter, so that the pole can not slip in the crotches.

Into this pole studs are mortised, which extend as low as necessary. Boards are nailed upon these studs, upon the up-stream side, and lapped so as not to catch the "drift," whatever it may be. When high water comes, this hanging gate will float upon the stream, the pole turning in the crotches, which must of course be somewhat higher than the floods can ever reach.

D. M. Hays, Fayette Co., Ohio, sends the description of one, fig. 3, which he calls a "flood-gate." He says: "My plan is the best I can get after long experience. If built of sound timber it will stand 15 years, as I have already tested." It is all of hewn timber; the posts 8 by 8 inches, and of length sufficient to rise above the floods, are set and braced in mud-sills (12 x 20 inches) not shown in the cut. The cross-beam, or plate, is mortised upon the posts, strengthened to prevent sagging, by a king-post (which is attached by a stirrup), and braces. The two gates are suspended independently, from the cross-beam, and are constructed on the same principle as the one in the second plan described, with respect to lap of boards, etc.

Fig. 4 illustrates the plan used by C. G. Siewers, of Ohio. It is immovable, and is adapted to a brook or "dry branch" liable to flood, after heavy rains. It consists of a log laid upon stones at a proper height above the bed of the brook, and against two strong posts. Upon this rails are laid, their ends bedded in the ground and fastened with stones. Stakes are driven on the sides to prevent pigs getting through. This is recommended as useful in filling up ravines, for much drift is caught which would otherwise be washed down to a lower point, and the bed of the stream is thus gradually raised.

The plan submitted by Mr. P. A. Bettersnout, Switzerland Co., Ind., is similar to this in object and principle, and consists of a

timber, fig. 5, built into two stone piers. Rails are laid in the bottom of the stream and mortised firmly into or fastened against this cross-timber.

No one kind of water-gap can be recommended as adapted to general use, but each of these kinds, and perhaps others, may be used under different circumstances. The hanging gates, unless they are quite heavy, may be swung by hogs so that they can get through, and the bed of the stream becomes nearly dry. This may be prevented by a stake driven on the upper stream side, to prevent the gate swinging in that direction, and a row of stakes to prevent the approach of the hogs on the same side. (1864)

Wind Mills

There are doubtless very many locations, where power may be obtained for farm purposes by the erection of wind mills. Many flour mills in this country, as well as in Europe, have long been driven by this power, and the fact of their continued use, is demonstrative of their utility. We are constantly inquired of as to the best form. We have also some half-a-dozen commendatory communications on the subject, but all of them, or all but one, are from individuals, directly or indirectly interested in some particular patent.

We are very desirous of supplying our readers with correct information on the subject, for it is needed; but we frankly confess that we are not yet able to give a well-grounded opinion, and in this, as in other instances, we prefer to be silent, until able to judge and speak intelligently—even at the risk of losing an editor's reputation for knowing everything. We hardly know how to get at this matter without spending a month or two, or more, and traveling several thousands of miles to thoroughly and carefully examine the various mills constructed, and in operation in different parts

of the country. If we can get at the matter in any way during the year, we shall certainly do so. (1859)

Wind Mills

The force of wind may be usefully applied by almost every farmer, as it is a universal agent, possessing in this respect great advantages over water power, of which very few farms enjoy the privilege.

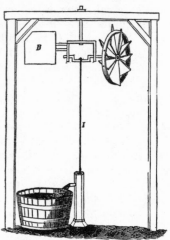

Fig.1.—A, *windmill*; B. *vane*; I, *pump-rod*.

Wind may be applied to various purposes, such as sawing wood by the aid of a circular saw, turning grindstones, and particularly in pumping water. One of the best contrivances for pumping is represented by fig. 1, where *A* is the circular windmill, with a number of sails set obliquely to the direction of the wind, and always kept facing it by means of the vane, *B*. The crank of the windmill, during its revolutions, works the pump rod, *I*, and raises the water from the well beneath. In whatever direction the wind may blow, the pump will continue working. The pump rod, to work steadily, must be immediately under the iron rod on which the vane turns. If the diameter of the windmill is four feet, it will set the mill in motion even with a light breeze, and with a brisk wind will perform

FIG. 2

the labor of a man. Such a machine will pump the water needed by a large herd of cattle, and it may be placed on the top of a barn, with a covering, to which may be given the architectural effect of a tower or a cupola, as shown in fig. 2. (1873)

Wind Mill Sails

In all wind mills, it is important that the sails should have the right degree of inclination to the direction of the wind. If they were to remain motionless, the angle would be different from that in practice. They should more nearly face the wind; and as the ends of the sails sweep around through a greater distance and faster, they should present a flatter surface than the parts nearer the center. The sails should, therefore, have a twist given them, so that the parts nearest the center may form an angle of about 68 degrees with the wind, the middle about 72 degrees, and the tips about 83 degrees.

In order to produce the greatest effect, it is necessary to give the sails a proper velocity as compared with the velocity of the wind. If they were entirely unloaded, the extremities would move faster than the wind, in consequence of its action on the other parts. The most useful effect, is produced when the ends move about as fast as the wind, or about two thirds the velocity of the average surface.

The most useful wind is one that moves at the rate of 8 to 20 miles per hour, or with an average pressure of about 1 pound on a square foot. In large wind mills, the sails must be lessened when the wind is stronger than this, to prevent the arms from being broken; and if much stronger, it is unsafe to spread any, or to run them.

Wind mills, for farm purposes, are apt to be broken by storms. To remedy this difficulty, springs have been attached to the sheet-iron sails, so that the strong wind shall turn them around with the edges to its course, and thus lessen the force of action. (1873)

Halladay's Wind Mill

Steam, horse, and water power, have been variously used for driving stationary machinery. The two former require the expenditure of fuel or feed, and the latter does not exist on many farms, and can be only occasionally used. But there is another, and universal power—found on every part of every single farm in the world—and sweeping over all with a strength of thousands of horses—which has been very little used for farm purposes. This is wind.

The great difficulty in the way of the general use of wind power is its unsteadiness. Common wind mills of much size cannot be run in any weather of a tempestuous charac-

ter, and much care is needed in regulating, rendering its ordinary use impracticable. Small wind mills, not over four feet in diameter, have been successfully applied to the pumping of water, where the wells were not deep, but unless well made, even these are liable to be broken by strong blasts.

These difficulties have been very successfully overcome by Halladay's Self-Regulating wind mill, invented by Daniel Halladay, of Ellington, and manufactured by the Halladay Windmill Company, of South Coventry, Ct. The self-regulating part is not unlike that of the governor of a steam engine. When a strong wind drives the mill too fast, the excess of water driven by a forcing pump against a piston, is made by a set of rods and levers to turn the edges of the sails more against the wind, and when the wind subsides, the same cause restores them to their former position.

These wind mills are chiefly intended for farmers, and may be applied to various other purposes besides pumping water. It is now about two years since their manufacture was commenced, and we have heard of none out of the many which have been erected, that have been blown down or injured by the wind. The cost of the smaller size is $75. (1873)

Wind Power—Wind Engines

The cheapest motive power in existence is the force of the wind. It can be utilized without preparation; no reservoirs, dams, or flumes are needed to apply it to our machinery, and the proper engine alone is to be provided. In some countries wind power is extensively used. The traveler in Europe scarcely loses sight of a windmill in his journeys, and in places the landscape is thickly dotted with them. Substantial grist mills, which have faced the breezes for centuries, still wave their arms and promise to do so for centuries more. Much pumping and drawing is done by these mills, and thousands of acres are either watered by irrigation or dried by drainage, and rendered valuable and productive by their help.

A few years ago a wind mill was an unusual sight in this country, except in the very oldest portions. We were not a sufficiently settled people, and did not remain long enough in one place to make it profitable to build such substantial mills as have been so long in use in other countries; we needed cheaper and more quickly constructed mills. Those which we could then procure, were not satisfactory, they were slightly built, and were not able to take care of themselves when the breeze became a gale or a hurricane. Recently our mechanics have turned their attention to wind engines, and great improvements have been made in their

construction. We have now a choice of several kinds of them, all of them useful, but differing chiefly in their degree of adaptation to varying circumstances.

At the recent Illinois State Fair there were no less than 13 different wind engines on exhibition, from the small one, 8 feet in diameter, costing but $100, of half a horse-power, and fitted for pumping stock water or churning, up to those of 30 or 40 horse-power, costing $3,000, and able to run a grist mill or a woolen factory. Between these extremes there are a number of mills capable of adaptation to almost every purpose for which power is needed on the farm or in the workshop.

A mill 22 feet in diameter, costing about $500, has a power of five horses; a two-horsepower mill is about 16 feet in diameter, and costs about $325. This cost is less than that of a steam engine, and a wind engine needs neither fuel nor skilled attendance. Neither is there danger of fire or explosion from accident or carelessness. The wind engines are now made self-regulating, and in a sudden storm close themselves. They are also made to change their position as the wind changes, facing the wind at all times.

With these engines one may saw wood or lumber, thrash, pump, hoist hay or straw with the hay fork, shell corn, grind or cut feed, plane lumber, make sash or doors, or run any machinery whatever. There is but one drawback, when the wind stops the mill stops. For work that may be done when it is convenient to do it, as most of the mechanical work on a farm is done, these engines are exactly what is wanted. On the Western prairies, and almost everywhere, except in sheltered valleys in the East, we have wind power enough and to spare, which offers to use a power that is practically illimitable, and the means of utilizing this power is cheaply given to us in the numerous excellent wind engines now manufactured. In fact so cheaply can these mills be procured, that it will not pay for any person to spend his time in making one, although he may be a sufficiently good mechanic to do it. Where there are several nearly perfect machines, we can not undertake to way which is the best. (1875)

Windmill for a Farm Shop

Mr. Gustav A. Michael, Montgomery Co., Pa., writes: "In answer to your request of some months ago, I send a few rough sketches of a windmill which has been adjusted for doing shop work, as boring fence posts, etc. I constructed the mill myself, and at a cost of not over five dollars.

The windmill wheel was made as follows: I selected one of the large wheels from an old horse-power, and removed the rim so that only the hub portion and the spokes were left. These were bevelled off, and the wings or fans of the mill wheel were fastened to them by means of screws. The wings are 4 feet long, 20 inches broad at the upper ends, and 6 inches at the bottom, so that the wheel measures 8 feet in diameter. The gearing house, fig. 2, is made of 2 by 6 inch oak plank mortised together. Two large holes, for the adjustment of the shafts, the lower and

larger, 5 inches in diameter, and the upper 3 inches, are cut somewhat to one side of the middle of the gearing house, so that the windmill may balance more perfectly upon the shaft, and therefore turn easily to the wind. The shaft is a one-inch iron rod obtained from an old mowing machine. There are two cog-wheels in the gearing house, by means of which the power and motion are transmitted from the windmill wheel to the shop below.

The beam upon which the turning shaft rests runs through the middle of the upper portion of the shop, and is of oak 5 by 5 inches in diameter. There is a gearing wheel on the lower end of the shaft that is 11 inches in diameter, and connects with another 2 feet in diameter, which is fastened to the upper portion of the shaft bearing the auger, or other tool to be used. The motion and general ac-

tion of the mill is governed by a brake. There is also a lever by means of which the auger, etc., is raised or lowered when at work. When posts are being bored they are placed upon a carriage provided with four wheels, which can be readily moved along as desired. I intend to arrange this mill as soon as time permits to give power for sawing wood, grinding tools, pumping water, etc., etc."

The windmill is shown in position, with the interior of the shop in view, in fig. 1; the

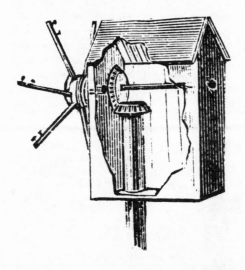

Fig. 2.—GEARING HOUSE.

Fig. 3.—CARRIAGE FOR HOLDING POSTS.

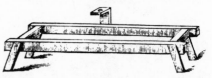

Fig 4.—THE WORK-STAND.

gearing house with the arrangement for shafts is shown in fig. 2, and the carriage and stand for holding the work are given in figs. 3 and 4 respectively.

Every farmer, from necessity, is more or less a machinist, and we have advocated from time to time that all should, if possible, have a shop in which they can do their work of repairing farm implements, and even of making new ones as they are wanted. There is no doubt but that the wind could be utilized at a trifling cost as a source of power in doing much of the important work connected with a farm shop. If one can afford the outlay, and it is not very great, required to purchase any of the many windmills now offered by the makers, he would no doubt have a more satisfactory wind engine than any he could make. But, as our correspondent shows, those who do not care to buy a windmill, and there are many such, need not be deprived of the use of wind power. (1881)

Improvement in Wind Engines

The power of the wind is much greater than is generally suspected. If the air were visible it would be seen in motion very much like water, the current smoothly flowing, when no obstructions interfered, but whirling, eddying, and irregular in its force, as obstacles to a free movement deflect the current into many directions. In making use of the wind as the cheapest available mechanical force the peculiarities of its motion must be taken into account.

Air possesses weight, and in moving against a fixed object its momentum exerts a force just as much as if it was a solid body that moved. The force exerted by the wind upon any stationary object is found by multiplying the square feet of surface opposed to the wind by the number of feet through which the wind moves in a second multiplied by itself, or squared and divided by .002288 or 2288/1000000; thus one of the 10-foot wind engines having 68 square feet of surface, in a wind moving 30 feet per second, or about 20 miles in an hour, which is what may be called a pleasant, brisk breeze, will receive a propelling force of 136 pounds, or more than equal to the amount of tractive force exerted by a horse moving at the rate of 3 miles in an hour. It is evident that the full utilization of this power of the wind depends very much upon the excellence of the machinery to which it is harnessed to do our work, so to speak; the more effective this may be, the greater amount of the force of the wind is turned to useful account. It is quite as important to investigate this point in choosing a wind-engine as it is in selecting a waterwheel. In calling attention to several excellent wind-engines we merely remark that for most of the work of the farm, dairy, and household, and for many mechanical works, a wind-engine is the most useful, economical, simple, and safe power that can be employed. With proper self-regulating devices it will work steadily night and day, without feeding, without watching, or other attention, needing only occasional oiling, to work on, while the owner sleeps, eats, or labors elsewhere. The engine is able to regulate itself, instinctively, as it were, to all the changes of the wind; turning out of the wind and stopping when it blows with excessive force, and turning in again and resuming work when the gale modulates. No other mechanical power can thus be left to regulate itself, except perhaps a waterwheel, and even this, in floods—which may be parallel in their effects to a gale—suspends its working.

The Eclipse windmill, made at Beloit, Wis., is the first that occurs to us. This made a wide reputation at the Centennial as the only mill there that was not damaged by the great storm of June 27th of that year. It is shown at work in fig. 1, and out of the wind at fig. 2. A side vane (shown at fig. 1)

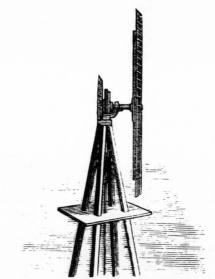

Fig. 2.—THE ECLIPSE OUT OF THE WIND.

round, one button working on the up stroke and the other on the down stroke, so that the wheel turns regularly and continuously. It is claimed that 120 revolutions per minute may be made in a 12-mile (an hour) wind. This enables the mill to churn, saw wood, run a lathe, cut and grind feed, etc., as easily as to pump water.

The Challenge windmill, made at Batavia, Ill., is a popular mill, and is both single and double. A double-headed mill, erected on the barn of the well known breeder of Hereford cattle, T. L. Miller of Beecher, Ill., is shown in fig. 4. This is a 30-foot mill, and cuts hay, shells and grinds corn, and pumps water for 300 cattle, 200 sheep, and 200 hogs; besides

doing the grinding for the neighbors.

Fig. 5 is the Myers' windmill, a rosette wheel with rudder, made at Salem, Ohio. The self-regulating device of this wheel is peculiar. The vane is parallel with the wheel, and lies flat when the mill is in gear. When the force of the wind overbalances the weight, this throws up the vane, and the wind

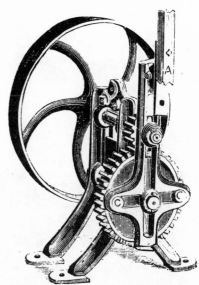

Fig. 3.—A MECHANICAL ATTACHMENT.

striking it, turns the wheel out of gear. When the wind abates, the vane is turned, and the wheel is brought into work again, as seen in the engraving.

At fig. 6 is shown the Perkins' mill, made at Mishawaka, Ind. This mill has been made since 1869, and is of the solid or rosette

Fig. 1.—THE ECLIPSE WINDMILL AT WORK.

receives the force of a storm and turns the mill edgewise to the wind. When the force abates a weighted arm brings the front to the wind again. A remarkable mechanical device, shown at fig. 3, is attached to this mill. It is one that changes an up and down movement to a rotary one. As the rod, A, moves up and down, the buttons, C, press on the inner rim of the gear wheel and carry it

Fig. 4.—A DOUBLE-HEADED MILL ON A BARN.

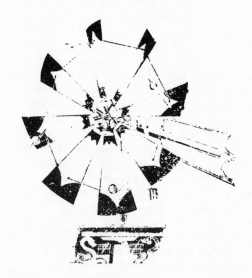

equalize the motion as the wind may change. The regulating apparatus consists of the balls or weights seen in the engraving, which govern the position of the sails according to their velocity.

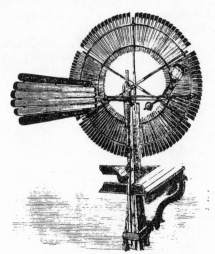

Fig. 5.—THE MYERS' WINDMILL.

form; it is also a self-governor. Another solid or "rosette" mill is the "I.X.L." made at Kalamazoo, Mich. (fig. 7). This is a self-governor, the wheel and rudder folding together in a wind too heavy to be safe. It is able to turn partly out of the wind and still run when the blow is not too heavy.

Fig. 8 is the Victor mill, made at New London, Ohio, which is moved by broad sails, so fitted as to turn on pivots and regulate and

The Stover wind engine, fig. 9, made at Freeport, Ill., is a "rosette" mill.

Fig. 6.—THE PERKINS' MILL.

Fig. 7.—THE "I. X. L." MILL.

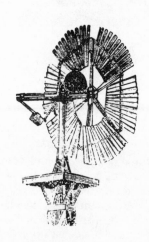

The Triumph engine, fig. 10, is a recent invention. (1882)

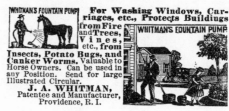
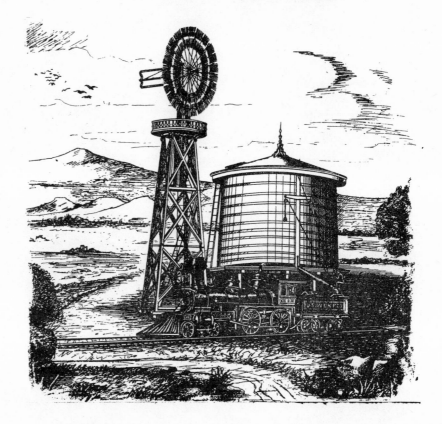

Index